Distinguished Dissertations

Springer-Verlag Berlin Heidelberg GmbH

John Bainbridge

Asynchronous System-on-Chip Interconnect

 Springer

Dr John Bainbridge
Department of Computer Science, University of Manchester, Manchester M13 9PL

Series Editor
Professor C.J. van Rijsbergen
Department of Computing Science, University of Glasgow, G12 8RZ, UK

British Library Cataloguing in Publication Data
Bainbridge, John
  Asynchronous system-on-chip interconnect. - (Distinguished
  dissertations)
  1.Asynchronous circuits 2.Integrated circuits
  3.Microcomputers - Buses
  I.Title
  621.3'815
  ISBN 978-1-4471-1112-2
Library of Congress Cataloging-in-Publication Data
A catalog record for this book is available from the Library of Congress.

Distinguished Dissertations ISBN 978-1-4471-1112-2

ISBN 978-1-4471-1112-2   ISBN 978-1-4471-0189-5 (eBook)
DOI 10.1007/978-1-4471-0189-5

a member of BertelsmannSpringer Science+Business Media GmbH
http://www.springer.co.uk

Preface

A shared system bus is a key feature of modern system-on-chip design methodologies. It allows the independent development of major macrocells which are then brought together in the final stages of development. The use of a synchronous bus in a synchronous design brings with it problems as a result of clock-skew across the chip and the use of many timing domains in a system. In an asynchronous system, the use of a synchronous bus would subvert many of the benefits offered by asynchronous logic such as reduced electromagnetic emissions.

This book was written as a doctoral thesis submitted in March 2000. It describes an asynchronous system-on-chip bus which offers a solution to such problems. Existing shared-bus techniques are re-investigated in the context of an asynchronous implementation, and a complete bus design is presented that was developed for use in an asynchronous subsystem of a mixed-synchrony chip. This chip was to form part of one of the first commercially available products to incorporate components that used asynchronous very large scale integration (VLSI) techniques, but was never productised.

The split-transfer primitive, often avoided or added as an optional extension by synchronous designers, is used as the basis for the chosen bus architecture. It offers a fine-grained interleaving of bus activity and a better bus availability than would an interlocked-transfer technique as found in many synchronous alternatives. This technique is viable in an asynchronous design because of the very low arbitration latency.

Simulation results (and fabricated prototype chips) show that the proposed architecture achieves a performance comparable with synchronous buses that use similar levels of resource, whilst maintaining the benefits of the asynchronous design style.

Since writing this thesis, work on asynchronous logic has been ongoing, and some of my suggestions for "future work" are being explored [7]. Further developments in Asynchronous System-on-Chip Interconnect and asynchronous logic design can be found online at our group World Wide Web pages which are permanently up to date:

http://www.cs.man.ac.uk/amulet

John Bainbridge
University of Manchester, UK
December 2001

Acknowledgements

Whilst working as a part of the AMULET group at the University of Manchester's Computer Science Department I have received support and inspiration from many of my colleagues, especially those involved directly with the AMULET3 project. Special thanks are due to a few individuals:

First and foremost of these is my supervisor, Prof. Steve Furber without whose valuable guidance and support this work would not have been possible. Many thanks Steve.

Dr Jim Garside and Dr Steve Temple have provided much advice on low level implementation issues and assistance with the CAD tools when problems arose, including a hack for the y2k bug of the Compass Design Tools. Steve also served as an interface to the commercial partners of the project, proof-read the book and manually laid out the top-level wiring of the chip.

Dr David Lloyd performed the unenviable task of converting my behavioural models written in an in-house language into a VHDL model for use by our commercial partner, and in the process raised many interesting issues for discussion.

Finally, I would like to acknowledge Dr Andrew Bardsley, Dr Phil Endecott and Dr Dave Gilbert for many useful discussions; Dr Doug Edwards and Prof. Ian Watson for introducing me to the fine arts of navigation and the downhill ride to the pub; and my parents for their timeless support whatever I choose to do.

Contents

List of Figures

List of Tables

1. Introduction

The major challenge that faces designers of System-on-Chip (SoC) integrated circuits is achieving the required functionality, performance and testability whilst minimising design cost and time to market. The key to achieving this goal is a design methodology that allows component reuse. Such methodologies have been an enabling factor in the success of vendors of intellectual property, such as ARM Ltd, who license designs of the same processor core macrocells to many competing semiconductor manufacturers.

The problem of design reuse has arisen at earlier stages in the development of electronic technology. A case in point is 8-bit microprocessor systems where the designer could pick-and-mix cards from many vendors' libraries, connecting the cards through a backplane. This approach only works if standard interfaces are widely used. Clocked buses were used for very simple systems, but asynchronous buses were preferred for high-performance or large-scale systems. In more recent times, the synchronous PCI bus [72] has become the dominant solution for board-level interconnection, driven largely by the x86 [46] PC processor market which provides a ready supply of both processors, peripheral cards and chips with interfaces for direct connection to the bus. However, asynchronous buses are still used, such as SCSI [77] in high-end disk subsystems and VME [87] and the IEEE FutureBus [34] at the rack level.

Now that integrated circuits can contain many millions of transistors on one chip [79], the emerging SoC industry must again address the problem of component reuse. This time, the components are macrocells which may be sold in both hard (i.e. physical layout) and soft (i.e. descriptions for synthesis) forms. The solution to the reuse problem is again to have common interfaces and a standardised interconnection network or bus.

Industry has so far adopted only synchronous macrocell buses, such as ARM's AMBA [2], as the interconnect solution, using a global clock to control the transfers. Restricting the "common interface" using a clock may be a short-sighted solution in the light of impending clock-skew problems and anticipated requirements for SoC designs to incorporate multiple clock domains and asynchronous macrocells.

This book presents the case for an asynchronous solution to SoC interconnect using a split-transfer system, allowing the seamless connection of asynchronous and synchronous components with multiple clock domains. The potential benefits include:

- support for both clocked and self-timed macrocells;
- improved macrocell reuse;
- zero power quiescent state;
- reduced electromagnetic emissions.

Supporting arguments for these claimed benefits are presented later.

As a concrete illustration of the feasibility of asynchronous interconnect, Chapter 8 describes the AMULET3H subsystem of a commercial telecommunications controller chip. At the heart of this subsystem is the Manchester Asynchronous Bus for Low Energy (MARBLE), a dual-channel split transfer bus that is summarised in Chapter 8.

The remainder of this chapter presents the merits of both synchronous and asynchronous design styles and concludes with an overview of the rest of the book.

1.1 Asynchronous design and its advantages

Asynchronous design, where timing is managed locally (as opposed to globally with a clock system as in synchronous design), was used in the early days of computers (before the days of VLSI technology) when machines were constructed from discrete components. With the arrival of the integrated circuit, the synchronous design paradigm gained popularity and became the dominant design style due to the simple, easy-to-check, one-sided timing constraint that it provides. However, in recent years asynchronous design has been reborn. Research and tool development in this area are still advancing, spurred on by the following advantages that asynchronous design offers.

1.1.1 Avoidance of clock-skew

Synchronous design methodologies use a global clock signal to regulate operation, with all state changes in the circuit occurring when the clock signal changes level. As feature sizes decrease and integration levels increase, the physical delays along wires in a chip are becoming more significant, causing different parts of the circuit to observe the same signal transition at different times. If the affected signal is a clock, then this time difference, known as clock-skew, limits the maximum frequency of operation of a synchronous circuit.

Through careful engineering of the clock distribution network, it is possible to mitigate the clock-skew problem, but solutions such as balanced clock trees [88] are expensive in silicon area and power consumption and require extensive delay modelling and simulation [27].

The absence of a global clock in an asynchronous circuit avoids the problems of clock-skew and the complexity (and design time) of the clock distribution network.

1.1.2 Low power

Power consumption is important in many embedded systems where battery life is at a premium. In larger, higher performance, systems power consumption affects the packaging cost of the system due to the need both to supply the energy onto the chip and to remove the heat generated. With some recent integrated circuits (ICs) dissipating many watts and processor data sheets now featuring sections on thermal management [1,47] this has become a significant problem.

Asynchronous design can reduce power consumption by avoiding two of the problems of synchronous design:

- all parts of a synchronous design are clocked, even if they perform no useful function;
- the clock line itself is a heavy load, requiring large drivers, and a significant amount of power is wasted just in driving the clock line.

There are synchronous solutions to these problems, such as clock-gating, but the solutions are complex and the problems can often be avoided with no extra effort or complexity when using asynchronous design.

1.1.3 Improved electro-magnetic compatibility (EMC)

The global synchronisation of a clocked design causes much of the activity in the circuit to occur at the same instant in time. This concentrates the radiated energy emissions of the circuit at the harmonic frequencies of the clock. Synchronous design approaches to spreading this radiated energy across the spectrum, such as varying the clock period, are complex to implement and affect the performance of the system since the clock period can only be made longer (not shorter) than the minimum for safe operation of the circuit.

Asynchronous circuits produce broadband distributed interference spread across the entire spectrum, as confirmed by field strength measurements of the energy radiated by the AMULET2e microprocessor [33]. This can be a significant advantage in systems which use radio communication where interference must be minimised.

An asynchronous macrocell bus, even when connecting entirely synchronous macrocells, should still offer EMC advantages in that it allows an arbitrary phase difference between the clocks used in different regions of the chip, giving rise to some cancellation of the radiated electromagnetic fields.

1.1.4 Modularity

The performance of an asynchronous design can be improved by modifying only the most active parts of the circuit, the only constraint being that the communication protocol used on the interface to the module must still be obeyed. In contrast, for a synchronous design, improved performance can often only be achieved by increasing the global clock frequency which will usually require most of the design to be reimplemented.

In the context of SoC design, the implications of this are enormous. Consider a library of synchronous components, designed to operate at a given clock frequency, probably determined by one system design. Later, a different system is designed that uses a processor and DMA controller with a higher global clock frequency for performance reasons but requires the use of some of the peripherals in the library, for example an ISDN interface. The ISDN interface is still limited in its throughput to the ISDN data-rate, but either must be interfaced to the higher clock frequency using multiple clock domains in the one system, or must be redesigned to operate at the higher frequency. If the system and the library had been asynchronous, the same interface would be usable, even with the higher throughput components, without any modification whatsoever.

1.1.5 Better than worst-case performance

In a synchronous system the minimum clock period must be chosen to accommodate the combination of the worst-case variations in:

- power supply voltage;
- temperature;
- transistor speed (which varies due to variations in the processing of the silicon);
- data-dependent evaluation time, e.g. a ripple carry adder can complete an addition with a small carry propagation distance faster than one with a long carry propagation distance.

Typically, the worst case combination is encountered very infrequently, and so an asynchronous circuit that is not restricted by having to run at a fixed clock frequency can achieve better than the worst-case performance whenever all the worst-case conditions do not coincide.

1.2 Disadvantages of asynchronous design

Asynchronous design has a number of disadvantages compared to synchronous design which may account for the apparent unwillingness of industry to adopt such techniques.

1.2.1 Complexity

The clocked design paradigm has one simple fundamental rule; *every processing stage must complete its activity in less than the duration of the clock period*. Asynchronous design requires extra hardware to allow each block to perform local synchronisations to pass data to other blocks. Furthermore, to exploit data-dependent evaluation times, extra completion-detection logic is necessary. This added complexity results in larger circuits and a more difficult design process.

1.2.2 Deadlock

Control logic designed using an asynchronous design technique is likely to deadlock if an event is either lost or incorrectly introduced, for example as a result of noise or ionising radiation. Synchronous control circuits offer better tolerance of such problems where, for example, the extra event may cause an incorrect output, but will not normally cause a complete system deadlock. Of course in some systems neither alternative can be tolerated.

1.2.3 Verification

Verification of synchronous designs requires the checking of the static timing constraint imposed by the clock and of the logical functionality of each module. For an asynchronous design, verification is difficult due to the non-deterministic behaviour of arbiter elements, and deadlock is not easy to detect without exhaustive state space exploration. Formal techniques for asynchronous circuit design [13] may assist in this area.

1.2.4 Testability

Testing for fabrication faults in asynchronous systems is a major obstacle due to the non-deterministic behaviour of arbiter elements. This problem also affects synchronous designs which often have arbitration or synchronizers at their periphery where there is also non determinism and the additional problem that metastability is (incorrectly) assumed to resolve within a clock period. Furthermore, scan-path techniques are more difficult to apply [68] in an asynchronous design than in an equivalent synchronous design.

1.2.5 "It's not synchronous"

Synchronous design techniques are widely used and have been taught in universities for over two decades. Most designers are thus not familiar with asynchronous design techniques and the benefits they offer. Clocks are so ingrained in the industry that until asynchronous design techniques offer major advantages over synchronous approaches, asynchronous design may never become more than a niche activity.

1.3 Book overview

This chapter has made a case for asynchronous SoC interconnect and highlighted the pros and cons of asynchronous VLSI design. The remainder of the book is arranged as follows.

Chapter 2 presents the fundamentals of asynchronous VLSI design showing the timing models in common use. Summaries of a number of example designs, from a low-power pager based upon an asynchronous microcontroller to high-performance asynchronous microprocessors, show the commercial readiness of asynchronous design.

Chapter 3 discusses the interconnection requirements of computer systems, with a discussion of the techniques used in large networks, at the system-board level and in on-chip systems.

Chapters 4, 5, 6 and 7 investigate the issues involved in designing an asynchronous macrocell bus, at four levels as shown in Figure 1.1.

This figure shows how a simple interface can be presented to the macrocell using two channels, one carrying commands (from an initiator to a target) and one carrying

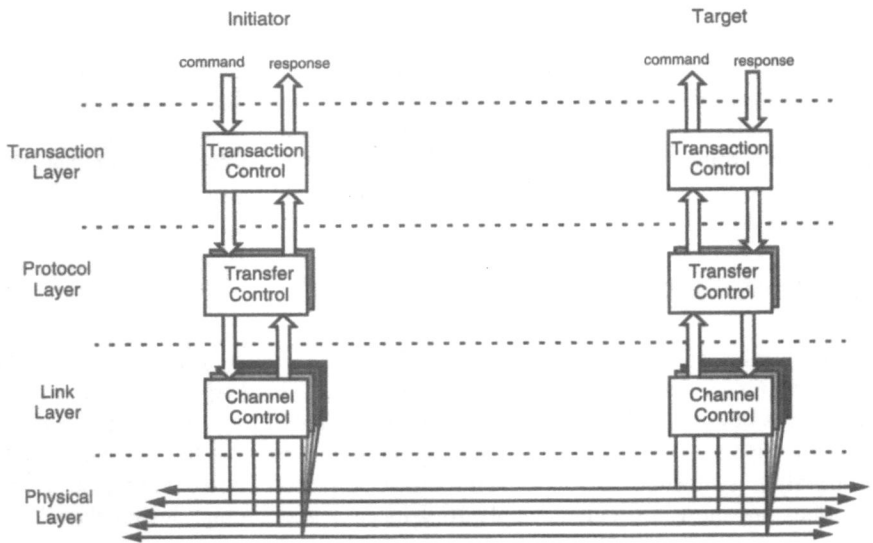

Figure 1.1: Bus interface modules

responses (from a target to the initiator). The bus-interface consists of components at
four levels, each level building more functionality on top of the lower level:

- A shared bus is in essence a group of wires with a fixed protocol ensuring that
  they will not be driven by multiple drivers at the same time. The wires
  themselves form the lowest level of the hierarchy presented in Figure 1.1, the
  physical layer. Important parameters of long wires on submicron VLSI
  technology include the length, width and separation, all of which affect the
  speed of signal propagation and the coupling between the wires. These issues
  are discussed in Chapter 4.

- An asynchronous multipoint channel that can have many senders and many
  receivers forms the next level of the hierarchy, the link layer, as discussed in
  Chapter 5. This chapter looks at grouping a set of wires and imposing a
  signalling protocol upon them so that ordered communications between
  senders and receivers can be performed. Issues addressed include arbitration
  for access to the channel, channel signalling protocols, address decoding and
  signalling and data hand-over. Centralisation versus distribution of the control
  for the channel is also discussed. Finally, asynchronous multipoint channel
  controller implementations are shown that allow the connection of a non-
  multipoint channel to the multipoint channel.

- For performance reasons a macrocell bus will usually use more than one
  multipoint channel. The middle level of the hierarchy, the protocol layer,
  defines the bus-transfer protocol and how the macrocell bus is built upon the
  link layer. Chapter 6 investigates the issues involved at this level where transfer
  phases must be mapped onto cycles on the channels provided by the lower-

level link layer and measures must be taken to ensure that activity on different channels that belongs to the same transfer is routed between the same two devices.

- The previous three levels provide for transfers between initiators and targets. The transaction layer, addressed in Chapter 7, is responsible for mapping the command and response channel packets passing between the bus-interface and the client device onto individual bus transfers. Simple transaction layers only permit one outstanding transfer across the bus at any instant, whereas more complex implementations allow a greater number, and hence deeper pipelining of the entire system.

As a concrete example of the viability of asynchronous SoC interconnect, Chapter 8 presents MARBLE, a complete specification for a dual-channel asynchronous bus using a split transfer protocol, in the context of the AMULET3H telecommunications subsystem, thus bringing together the issues discussed in the earlier chapters. Full details of both initiator and target interfaces are included with descriptions of features other than the protocol, such as support for burst optimisations.

Chapter 9 presents an evaluation of the MARBLE bus architecture. Performance metrics analysed include the bandwidth and latency of the system. The delays inherent in the different parts of the bus system are also investigated to highlight potential areas for improvement.

Finally, Chapter 10 draws conclusions about how MARBLE demonstrates the advantages and disadvantages of an asynchronous macrocell interconnection strategy in addition to suggesting future expectations and possibilities for the use of asynchronous SoC interconnect.

1.4 Publications

The following papers, based on the work presented in this book, have been published or submitted for publication:

- Asynchronous macrocell interconnect using MARBLE [5]
  (ASYNC 98 Conference);
- MARBLE: A proposed asynchronous system level bus [8]
  (2nd UK Asynchronous Forum);
- Bridging between MARBLE and a clocked peripheral bus [9]
  (4th UK Asynchronous Forum);
- Multi-way arbitration [10]
  (5th UK Asynchronous Forum)
- Crosstalk analysis for AMULET3/MARBLE [11]
  (5th UK Asynchronous Forum)
- MARBLE: An asynchronous on-chip macrocell bus [6]
  (Microprocessors and Microsystems, Vol 24, No. 4 August 2000)
- AMULET3i - an Asynchronous System-on-Chip [35]
  (ASYNC2000 Conference)

2. Asynchronous Design

This chapter provides an introduction to asynchronous design. The information presented here is intended to set the context for the overlap of two themes: asynchronous design and SoC interconnect, in the form of an asynchronous macrocell bus. Further details on all aspects of asynchronous design are available elsewhere [85].

2.1 Introduction

Fifty years ago, computer pioneers designed digital computers based upon thermionic valves [89,90], and systems such as the Ferranti Mark 1 [32] constructed on large racks with custom inter-rack connections occupied large rooms or buildings. The principles of binary digital design were the same then as they are now:

- two values, 0 and 1, represent information and they are represented as distinct signal values (most commonly voltages);
- signals must only be sampled or observed when in one of these two distinct states.

Much of the methodology of digital design is a consequence of the latter requirement, imposed upon the designer because the *digital circuit* is really an analogue circuit approximating a digital behaviour. The approximation breaks down at the point when a circuit switches state from 0 to 1 or from 1 to 0, which takes a significant time as the signal passes through the analogue space between the two digital threshold voltages. If the value is sampled during this time, which appears as a delay in the digital model, its value is unpredictable and may give unexpected behaviour (a phenomenon known as a hazard). Digital design methodologies differ in how they indicate when the signals are in a stable 0 or 1 state (and are thus ready to be sampled reliably):

- The synchronous (clocked) methodology globally distributes a timing signal to all parts of the circuit. Transitions (rising and/or falling depending on the design) on this clock line indicate moments at which the data signals are stable.
- The asynchronous (self-timed) methodology utilises the passing of time, as indicated by local matched delay lines, to indicate when the data signals are stable, or encodes the timing information in the data line activity itself.

The advent of the transistor meant that much larger, more reliable, systems became feasible and design methods advanced to meet the needs to handle greater complexity. Systems were constructed from racks using a backplane hosting many daughter cards, each with a standard interface, allowing extra functionality to be provided by adding more cards to the system. During this period both synchronous and asynchronous design styles were commonplace, the latter using techniques based on the use of local-clocks and delays based upon multiples of the clock period. Circuits of this era typically operated under what is known as the bounded delay model, where the delays in both the gates and wires are assumed to be within a bounded range (as opposed to the unbounded delay model where the delay can be any finite value). This approach was somewhat ad hoc, and when the VLSI era arrived, the more disciplined synchronous design style dominated.

The VLSI era has allowed whole system components to be constructed on a chip, requiring fewer boards and racks per system and allowing higher performance. Backplane buses advanced in complexity to meet these performance requirements.

Over the past two decades, research into asynchronous design has concentrated on finding more disciplined approaches which can challenge clock-based design in offering a reliable basis for VLSI design. Often unbounded delay assumptions are used, which guarantee that a circuit will always operate correctly under any distribution of delay amongst the gates and wires within the circuit.

Currently, asynchronous design methods are well developed (as shown later in this chapter), and whole computer systems can be constructed as a single integrated circuit using either synchronous or asynchronous methods. Now, the interconnect and reuse problems that affect designers at the circuit-board and rack level also affect the VLSI designer. Again the solution is to have a standardised interface allowing modules to be mixed-and-matched. This interface is the macrocell bus.

2.2 Asynchronous design

Fundamental to the understanding of asynchronous design is a familiarity with the assumptions commonly made regarding the delays in the gates and wires within a circuit and the mode in which the circuit operates. The two common delay models, bounded-delay and unbounded-delay, were introduced above. The bounded delay model was commonly used in the early days of asynchronous design, and is still used in some backplane level interconnection schemes such as the SCSI bus [77] where part of the protocol is based upon known, fixed delays. Current asynchronous VLSI designs and research efforts use the unbounded delay model for the implementation of state-machines and controllers since it leads to circuits that will always operate correctly whatever the distribution of delays. It separates delay management from the correctness issue, allowing the functionality of the circuit to be more easily verified. The bounded-delay model is still commonly used for datapath components, however, since in this area it can lead to smaller implementations.

The following sections discuss other aspects of various asynchronous design methodologies.

2.2.1 Circuit classification

Within the unbounded delay model, there are various different design styles in common use, each with its own merits and problems. In order of increasing number of timing assumptions they are:

Delay-insensitive (DI) circuits
A circuit whose operation is independent of the delays in both circuit elements (gates) and wires is said to be delay-insensitive. Martin has shown that the range of true delay-insensitive circuits that can be implemented in CMOS is very restricted [56].

Quasi delay-insensitive (QDI) circuits
If the difference between signal propagation delays in the branches of a set of interconnecting wires is negligible compared to the delays of the gates connected to these branches then the wires are said to form an *isochronic fork* [80]. Circuits created using the DI design style augmented with the isochronic fork assumption are said to be quasi delay-insensitive (QDI).

Speed-independent (SI) circuits
If wire delays in a circuit are assumed to be zero (or, in practice, less than the minimum gate delay), and the circuit exhibits correct operation regardless of the delays in any circuit elements, then the circuit is said to be speed-independent. The assumption of zero wire delay is valid for small circuits.

2.2.2 The channel

In asynchronous design, data is passed between modules using a group of wires, collectively known as a channel. These channels are normally unidirectional point to point connections, and over the years, a number of different asynchronous VLSI channel implementations have been defined. In such channels, the data always flows in one direction between two devices:

- the *sender* is the device that delivers data onto the channel;
- the *receiver* is the device that accepts data from the channel.

Orthogonal to this classification is the concept of which end caused the transfer to occur:

- the *initiator* is the device that caused the transfer to occur;
- the *target* is the device that responds to the initiator.

Which device performs which function is determined by the protocol and transfer direction used on a given channel.

2.2.3 Signalling conventions

The transfer of information across a channel is negotiated between the sender and receiver using a signalling protocol. Every transfer features a *request (req)* action where the initiator starts a transfer, and an *acknowledge (ack)* action allowing the target to respond. These may occur on dedicated signalling wires, or may be implicit in the data-encoding used (as described below), but in either case, one event indicates data validity, and the other signals its acceptance and the readiness of the receiver to accept further data.

The flow of information relative to the request event determines whether the channel is classified as a *push* channel (where information flows in the same direction as the request) or a *pull* channel (where information flows in the same direction as the acknowledge). These two types of channel are illustrated in Figures 2.1a and 2.1b. Designers often speak of pushing or pulling data, thus implying the protocol used.

The request and acknowledge may be passed using one of the two protocols described below; either a 2-phase event signalling protocol (a non return-to-zero scheme) or a 4-phase level signalling protocol (a return-to-zero scheme). Conversion between the different protocols has been well documented elsewhere [53], with many types of latch controller documented for converting between the different 2-phase and 4-phase signalling protocols.

2-phase (transition) signalling

In the 2-phase signalling scheme, the level of the signal is unimportant; a transition carries information with rising edges equivalent to falling edges, each being interpreted as a signalling event. A push channel using the 2-phase signalling protocol thus passes data using a request signal transition, and acknowledges its receipt with an acknowledge signal transition. Figures 2.1c and 2.1d illustrate the push and pull data-validity schemes for the 2-phase signalling protocol.

Proponents of the 2-phase design style try to use the lack of a return-to-zero phase to achieve higher performance and lower power circuits.

4-phase (level) signalling

The 4-phase signalling protocol uses the level of the signalling wires to indicate the validity of data and its acceptance by the receiver. When this signalling scheme is used to pass the request and acknowledge timing information on a channel, a return-to-zero phase is necessary so that the channel signalling system ends up in the same state after a transfer as it was in before the transfer. This scheme thus uses twice as many signalling edges per transfer than its 2-phase counterpart. Push and pull variants of the 4-phase signalling protocol are shown in Figures 2.1e and 2.1f.

4-phase control circuits are often simpler than those of the equivalent 2-phase system because the signalling lines can be used to drive level-controlled latches and the like directly.

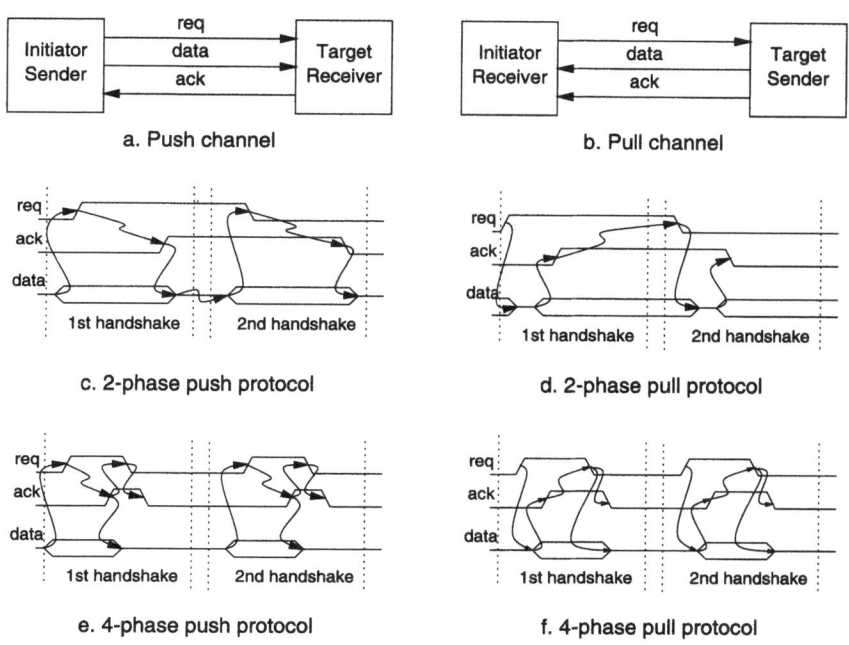

Figure 2.1: Channel signalling protocols

2.2.4 Data representation

A further dimension in asynchronous design is the choice of encoding scheme used for data representation where the designer must choose between a single-rail, dual-rail, 1-hot or other more complex N-of-M scheme. These alternatives are discussed in the following sections.

Single-rail encoding

Single-rail encoding [66] uses one wire for each bit of information. The voltage level of the signal represents either a logic 1 or a logic 0 (typically Vdd and Vss respectively for CMOS technology). This encoding is the same as that conventionally used in synchronous designs. Timing information is passed on separate request and acknowledge lines which allow the sender to indicate the availability of data and the receiver to indicate its readiness to accept more new data. This scheme is also known as the *bundled-data* approach. All single-rail encoding schemes contain inherent timing assumptions in that the delay in the signal line indicating data readiness must be no less than the delay in the corresponding data path.

Single-rail design is popular, mainly because its area requirements are similar to those of synchronous design, as is the construction of any arithmetic components using this scheme.

Dual-rail encoding

Dual-rail circuits [84] use two wires to represent each bit of information. Each transfer will involve activity on only one of the two wires for each bit, and a dual-rail circuit thus uses 2n signals to represent n bits of information. Timing information is also implicit in the code, in that it is possible to determine when the entire data word is valid by detecting a level (for 4-phase signalling) or an event (for 2-phase signalling) on one of the two rails for every bit in the word. A separate signalling wire to convey data readiness is thus not necessary.

The 4-phase dual-rail data encoding is popular for the QDI design style but, as with all dual-rail techniques, it carries a significant area overhead in both the excess wiring and the large fan-in networks that it requires to detect an event on each pair of wires to determine when the word is complete and the next stage of processing can begin. As an illustration of this point, Figure 2.2 shows a circuit fragment suitable for detecting the presence of a valid word on a 4-bit datapath. In practice, it is also necessary to detect when all of the bit-lines have returned to zero, for which the AND gates must be replaced by Muller C-elements as described in section 2.2.5.

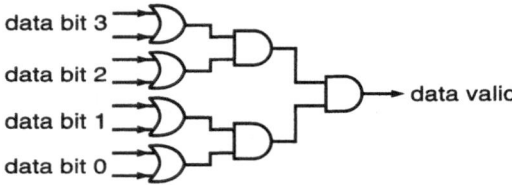

Figure 2.2: 4-bit dual-rail completion detection logic

One-hot encoding/Null Convention Logic (NCL)

One-hot circuits use 2^n signal lines to represent n bits of information, each line representing one n-bit code. One line can thus transmit n-bits of information and the associated timing information (data validity).

The Null Convention Logic (NCL) design style from Theseus Logic Inc. [31] uses a 1-hot encoding where each wire indicates either *data* or *null* combined with a 4-phase (return to zero) protocol. The scheme permits an arbitrary number of wires, e.g. decimal numbers could be represented directly using 10 wires, with only one wire in a group allowed to signal data at any one time. Typically, only two wires are used (as for dual-rail logic) where one wire is used to indicate true and the other false.

The NCL design style does not use Muller C-elements (as described in section 2.2.5), but instead uses threshold (majority) gates [73]. These gates give a high output when a specified number (or greater) of their inputs are high but, as for the Muller C-element, they feature hysteresis such that after the output has risen, it will not fall until all inputs have fallen.

For example, the gate shown in Figure 2.3 has five inputs and a threshold of 2. If any two (or more) of the inputs are high, then the output will switch to a high level. The output will then remain high until all the inputs are low.

Figure 2.3: 2-of-5 threshold gate

As with the dual-rail logic described above, NCL carries the timing information implicitly with the data, and again suffers from the large fan-in requirements of the logic necessary to detect when a complete word is present - a 4-bit data path requires a 4-of-8 threshold gate which can be realised as shown in Figure 2.4. This circuit has the same behaviour as that achieved if the AND gates of the circuit in Figure 2.2 are replaced by Muller C-elements, but is shown using threshold gates here to illustrate the NCL approach.

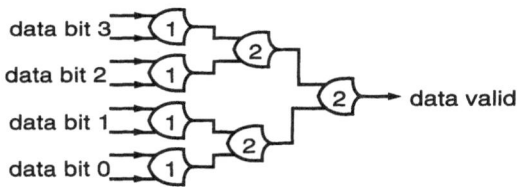

Figure 2.4: NCL 4-bit word completion detection logic

N-of-M encoding

Dual-rail encoding and one-hot encoding are examples of an N-of-M encoding scheme where N=1. Coded data systems using an N-of-M code where M>N operate correctly regardless of the distribution of delay in the wires or gates, and are thus said to be delay-insensitive [84].

More complex codes exist (where N>1) which use actions on more than one wire in a group to indicate one of a set of possible codes. These offer better utilisation of the available wires (for example a 2-of-7 code can transmit 4 bits of information over seven wires in a delay-insensitive manner), but result in larger arithmetic circuits and conversion between the coded form and a single-rail scheme is more expensive than for the 1-of-M codes.

2.2.5 The Muller C-element

The Muller C-element (often known as C-element or a C-gate) is commonly encountered in asynchronous VLSI design where it is used both for synchronising events and as a state-holding element. Figure 2.5a shows a symmetric 2-input C-element, its logic function and one possible pseudo-static CMOS implementation.

In a CMOS implementation of a symmetric C-element the n and p stacks are a reflection of each other. Asymmetric variants of the C-element have different structures for the n and p stacks, and thus some input signals may only affect either the rising or falling output transition, not both. Example asymmetric C-elements are shown in Figures 2.5b, 2.5c and 2.5d.

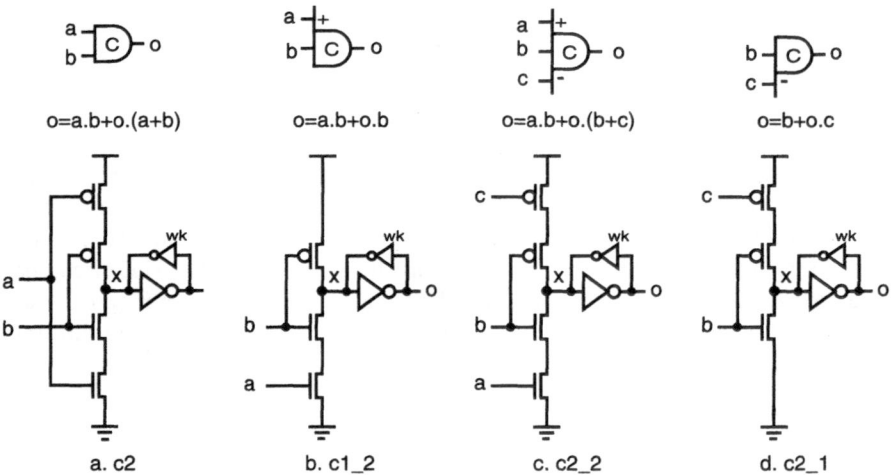

o=a.b+o.(a+b)   o=a.b+o.b   o=a.b+o.(b+c)   o=b+o.c

a. c2   b. c1_2   c. c2_2   d. c2_1

Figure 2.5: Muller C-elements, their function and their implementation

2.2.6 Specifications and automated circuit synthesis

A number of specification techniques are available to the asynchronous designer. Here
we summarise those for which a synthesis route is currently available. For small-scale
asynchronous designs, there are two classes of specification commonly used: state-
based and event-based, as described below. Unfortunately, these techniques do not
scale well, so they are only used for the construction of small modules. Larger designs
are then created from compositions of these small modules. Balsa [12] and Tangram
[81,82] are examples of such a synthesis approach based upon the syntax-directed
translation of a programming language based on CSP [40].

Event-based specification and synthesis
Petri Net [67] specifications describe the behaviour of systems in terms of sequences
of events, incorporating the concurrency and causality between the events.

Current event-based asynchronous circuit synthesis methodologies (including the
Petrify tool [21,22] used for some of the work presented in this book) are based upon
the foundations laid by Chu [17,18]. These use an interpreted Petri Net, known as a
Signal Transition Graph (STG), as the input specification with the transitions labelled
with signal names. In the STG notation a transition is labelled with either a "+"
(represents a rising signal), "-" (representing a falling signal) or a "~" (representing a
change in level). Dependencies and causalities are represented in the STG using the
notations shown in Figure 2.6. As an example of the STG specification style,
Figure 2.7 shows the specification of the 2-input Muller C-element described above
with inputs "a" and "b" and output "o". The dotted arcs show the behaviour of the
circuit's environment, and the solid arcs show the behaviour of the circuit (the C-
element in this case).

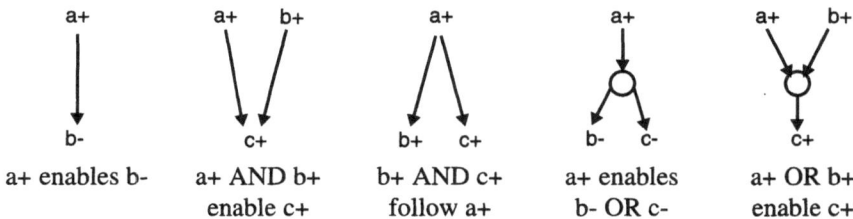

Figure 2.6: STG notation

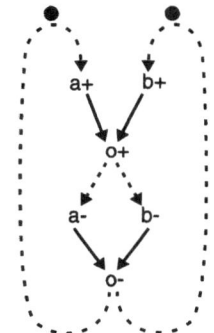

Figure 2.7: STG specification for a 2-input Muller C-element

State machine based specification and synthesis

Huffman state machines [43] are the classical asynchronous finite state machine, operating in fundamental mode. Burst-mode, as introduced by Stevens [19,26,74] and formalized by Nowick [62], and extended burst mode state-machines are a relaxation of the fundamental mode condition of only one input changing at a time. The MEAT tool [19] was the first burst-mode synthesis package, although this has been superseded by tools such as Minimalist [63].

2.2.7 Metastability, arbitration and synchronisation

Section 2.1 introduced the problems in modelling a digital system with an analogue approximation. Similar problems are encountered when a bistable system must determine an ordering of two asynchronous inputs that occur almost simultaneously. In this case, time is the continuously varying input, and during the period where the decision is made, the state of the output may be neither a 0 nor a 1, but somewhere in between. This situation is known as metastability [15]. Furthermore, making a reliable decision in a bounded time is fundamentally impossible [23,50,16,41] because a circuit with a finite gain-bandwidth product cannot resolve a continuously varying input into a discrete output in bounded time.

This is accommodated in a synchronous system by waiting for a predefined period, a clock cycle for example, before using the output from the bistable circuit. This act of sampling an asynchronous input at the instant when a transition on the other input (typically the clock in synchronous design) occurs is known as synchronisation.[1] There is always a chance of synchronisation failure in a synchronous system because the bistable circuit may still be metastable when its outputs are used in a subsequent clock period. Careful engineering to optimise the gain-bandwidth product of the bistable circuit can usually make this probability acceptably small but cannot eliminate the possibility of failure.

Asynchronous design uses the alternative approach of waiting until the bistable has resolved its output to a defined logical value, 0 or 1, before allowing the output to pass into the rest of the system. This act of determining which event came first is called arbitration. It can be achieved using an analogue filter attached to a bistable to create a *mutex* structure as proposed by Seitz [71]. A suitable CMOS implementation of Seitz' NMOS design is shown in Figure 2.8.

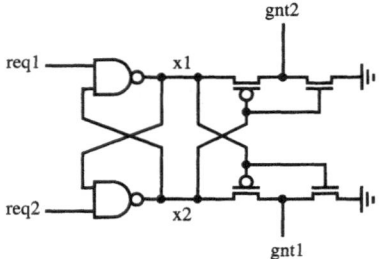

Figure 2.8: CMOS mutex implementation

Theoretically the arbitration can require an unbounded time, but in practice the probability of remaining in the metastable state for a long period is sufficiently small for it to be insignificant. If the occasional long delay can be tolerated then arbitration can be "failure free". Asynchronous arbitration thus incurs the average case delay whereas synchronisation always results in the worst case delay and a higher probability of failure.

2.2.8 Sutherland's micropipelines

In his 1988 Turing Award lecture, Ivan Sutherland introduced a framework for designing elastic asynchronous pipelines. The lecture, entitled "Micropipelines" [75], proposed a library of asynchronous modules (shown in Figure 2.9) with 2-phase, unidirectional bundled-data channel interfaces which could be interconnected to build larger asynchronous circuits.

The **or** function for events is provided by the exclusive-or (XOR) gate. This is also known as a *merge* because it allows two or more event streams to be merged into one.

---

1. Not to be confused with the synchronisation of two event streams as when using a C-element as the AND function in a 2-phase system as described in Section 2.2.8

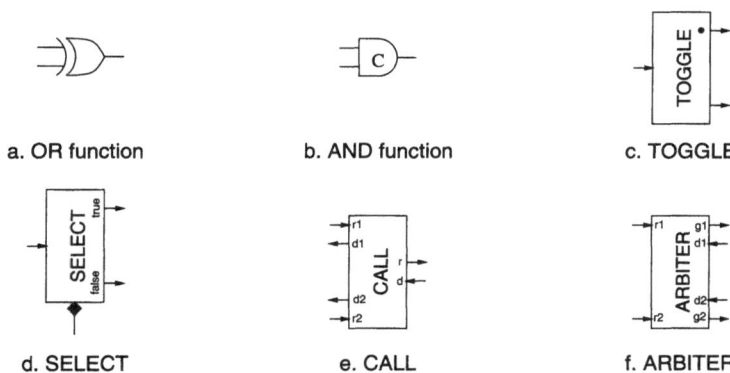

a. OR function   b. AND function   c. TOGGLE

d. SELECT   e. CALL   f. ARBITER

Figure 2.9: Micropipeline event control modules

For each transition on an input, a corresponding event will be seen on the output. For correct operation, input events must be well separated. In practice this means that each input event should be acknowledged by an output event before the next input event is applied.

The **and** function for events is provided by the Muller C-element. A transition will occur on the output only when there has been a transition on both of the inputs. The C-gate is also known as a *rendezvous* element because the first input event is held-up until it is joined by an event on the other input before being allowed to pass to the output (often referred to as a synchronisation of the two events).

The **toggle** steers events to its outputs alternately. The first event after initialisation is steered to the output marked with a dot, the next to the unmarked (blank) output and then the cycle repeats.

The **select** block again steers incoming events to one of its outputs. However here the output is selected by the level of the boolean select signal (indicated by the diamond in Figure 2.9d). The select signal must be set-up before the input event arrives, a similar requirement to the bundling constraint.

The **call** block allows two mutually exclusive processes to access a shared resource or procedure, much the same as a procedure call in software. The call routes input events on either r1 or r2 to the output r, and then routes the acknowledge/done event from the d input to either d1 or d2, dependent upon which input request event was routed previously. For correct operation the call block requires activity on the input channels to be mutually exclusive.

The **arbiter** provides arbitration between two contending asynchronous input streams (on r1 and r2). It handles metastability internally (as described in Section 2.2.7) whilst still presenting valid logic levels at its outputs (g1 and g2). Like a semaphore in software, it delays subsequent grants until it has received an event on the done wire (d1 or d2) corresponding to an earlier grant thus ensuring that there is no more than one outstanding grant.

2.2.9 Large asynchronous circuits

Asynchronous design is now achieving maturity with a number of groups having fabricated functional microprocessors. The following sections introduce the output of a few of the key players in the field of asynchronous design. The Philips pager described is the first commercially available product to incorporate an asynchronous VLSI chip.

AMULET

The AMULET group at the University of Manchester, UK, have produced a series of low power asynchronous reduced instruction set (RISC) microprocessors that execute the ARM instruction set. These processors all offer low power and good electromagnetic compatibility, with the latest release, AMULET3, delivering 100 Dhrystone 2.1 MIPS and a power efficiency of 780 MIPS/W. The AMULET processors use custom datapath layout and a single-rail bundled data signalling protocol.

AMULET1 [65] was a 2-phase implementation constructed to demonstrate the viability of the asynchronous micropipelined design style. Innovative asynchronous features of the AMULET1 included a simple ripple carry adder with a data-dependent propagation time and an instruction prefetch unit which had a non-deterministic (but bounded) prefetch depth beyond a branch.

Although functional, the final silicon implementation presented problems when using the 2-phase protocol at the system board level (there was no on-chip RAM), but the performance was within a factor of 2 of the equivalent synchronous ARM6 microprocessor.

AMULET2e [33] used a 4-phase protocol throughout and a reorganized pipeline allowing it to achieve a performance of 42 Dhrystone 2.1 MIPS (between an ARM 710 and ARM810). Power consumption was measured to be 150mW (giving a power efficiency of 280 MIPS/W). Major features of the AMULET2e included:

- A simple off-chip interface that relied upon an external delay line for its timing.
- A low-power halt mode where an idle-loop instruction causes a control event to be blocked. The stall eventually propagates though the processor, causing all activity to cease. The processor is re-awoken when an interrupt occurs.
- A branch target cache to reduce the frequency of pipeline flushes following branch instructions.
- A 4-KB on-chip cache which can also be used as a memory mapped RAM.

AMULET2e also used more carefully dimensioned FIFO buffers than AMULET1 because the AMULET1 design had shown that although deep asynchronous pipelines are easy to build, they can have an adverse effect on performance [33].

In a further bid for increased performance, the AMULET3 core uses a Harvard architecture and a five stage processing pipeline. A reorder buffer provides support for resolution of dependencies and late exceptions. This core has been developed in parallel with the MARBLE bus presented in this book, and the first version of both are included in the AMULET3H chip described further in Chapter 8. Coverage in more depth of the AMULET3 core architecture is also provided in Chapter 8.

TITAC-2

An asynchronous microprocessor based upon a modified MIPS R2000 instruction set has been designed by members of the Tokyo Institute of Technology. The TITAC-2 [60,78] processor uses a dual-rail encoding of data and a *Scalable Delay Insensitive* delay model. This model uses more relaxed delay assumptions than the QDI model, but is limited in application to small blocks, which are then connected using a QDI model.

Philips pager and 80C51 microcontroller

An asynchronous implementation of the 8-bit complex instruction set (CISC) 80C51 microcontroller and assorted peripherals have been synthesized using the Tangram toolkit by van Gageldonk [83]. The single-rail bundled-data asynchronous implementation of the microcontroller has a similar performance (4 MIPS) to its synchronous counterpart, but consumes only one third of the power at a cost of twice the silicon area.

This microcontroller has been used in production in the Philips Myna pager, where its low level of electromagnetic emissions offers a competitive advantage over an equivalent system based upon a synchronous microcontroller.

ASPRO

The ASPRO-216 [69] is a QDI 16-bit scalar RISC standard cell microprocessor from the Ecole Nationale Superieure Telecommunication (ENST) in Bretagne, France. In addition to the usual arithmetic and control instructions found in any RISC processor, its instruction set allows up to 64 custom instructions which can be used to drive additional "non-core" functional units. Instruction issue is performed in-order, but instructions are allowed to complete out of order, using a register locking mechanism to resolve dependencies.

2.3 Summary

Asynchronous design has reached the situation where a number of groups have successfully produced complex designs which are now beginning to appear in commercial applications. As such, asynchronous design now faces the same challenge as synchronous design: combining a number of macrocells to form a complete System-on-Chip design.

3. System Level Interconnect Principles

The increase in integration of functionality onto a single chip means that a greater demand is being placed on the on-chip interconnects, and so techniques that have been used in larger scale networks are now also being considered for on-chip use. This chapter discusses the basic concepts of a multipoint interconnect, with emphasis on the shared bus which is often favoured for its low hardware cost. Alternative approaches are discussed, but at present these are still expensive to implement for an on-chip interconnect when compared to a shared bus. Finally, a precis of three common synchronous SoC buses is provided.

3.1 Point-to-point communication paths

The simplest of link approaches, the serial point-to-point connection, is often used for long distance connections with the links implemented using wires, satellites or optical fibres. Slower (conventional wire-based) serial connections have many applications. The classical example of this category is the RS232 [70] protocol used for the serial ports of most modern microcomputers, and for the connection of terminal apparatus to older mainframe systems.

Where improved performance (throughput or latency) is required, often over shorter distances, parallel connections are commonplace. The parallel interconnection found between printers and their host computers is a prime example. On-chip buses are often latency critical, and hence use a parallel non-multiplexed approach with each bit of both the address and data flowing over separate wires.

3.2 Multipoint interconnect topology

Many systems require the interconnection of a number of devices. Depending on the performance requirements and the available hardware resource, one of the following alternatives will be used.

3.2.1 Shared buses

The shared bus is often favoured for multipoint connections because of its minimal hardware requirements, and there is a plethora of standards in this area. The same arguments over hardware requirements mean that the bus is also the favoured interconnection technique for on-chip use.

With a shared bus interconnection, many devices are connected to a group of wires (the bus), but only two devices are allowed to use the wires at any one time. The two devices are referred to as the initiator and the target (the same terminology as used for the asynchronous communication channel). The bus protocol is responsible for ensuring that the wires are used in an ordered manner, and for defining a mechanism for passing the active initiator and target roles between devices.

3.2.2 Star and ring networks

Alternatives to the shared-bus are the star or ring network. 10-base-T Ethernet is a commonly encountered example. These systems contain packet-switches at the centre of the star, or at each node of the ring, to route incoming packets to an outgoing link. The packet-switch can be implemented as a localised (hence very low load) shared-bus (maybe using a gated multiplexer rather than the tristate techniques usually associated with a bus) which can operate at very high frequencies.

3.2.3 Meshes

The ultimate in interconnectivity is the mesh approach, which provides multipoint interconnectivity using point-to-point links between nodes, the many links forming a mesh. As with the star and ring networks, the switches are in essence (fast) localised shared buses with a very low capacitive load.

As an example, two initiators can be connected to four targets using a group of buses arranged as shown in Figure 3.1. Here the address-decode is performed first to determine which of the horizontal buses to use. Then the merge function is performed on a horizontal bus-channel unique to the addressed target. The resulting cross-bar matrix presents exactly the same interface as that presented by a single channel, i.e. two initiators and four targets for the figure shown, but supports concurrent transactions when there is no conflict for a client.

3.3 Bus protocol issues

A bus is often preferred over the other topologies introduced above because of its lower implementation cost. This book addresses the design of an asynchronous shared system bus for use in on-chip systems. The features typically found in shared buses include the following.

3.3.1 Serial operation

Where large distances are involved, serial operation provides a cost advantage as it uses very few wires. A prime example of a commonly used serial bus is the 10-base-2 Ethernet which uses coaxial cable (two conductors) to interconnect a number of nodes and operates at 10 Mbit/s.

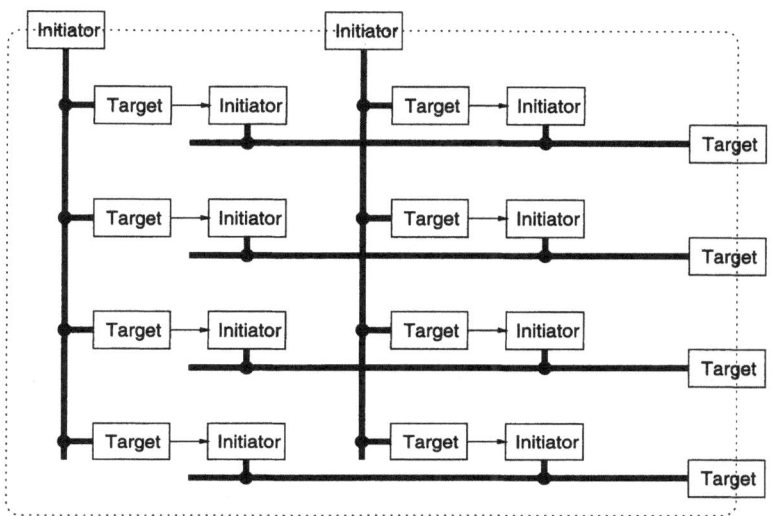

Figure 3.1: Fully interconnected bus-channel matrix

3.3.2 Multiplexed address/data lines

Parallel bus systems where the same signal lines are used to send first an address/command and then a data value are known as multiplexed address and data buses. These are usually found in backplane buses where connector pins are at a premium and serial buses cannot meet the required performance. The PCI [72] bus is an example of this approach.

3.3.3 Separate address and data lines

Where lower latency or greater bandwidth is required, the address and data are transferred over separate signal lines. Techniques such as pipelining of subsequent address and data transfers are very common in such buses.

Some variants of this approach use bidirectional tristate lines, with the direction changed between reads and writes, whereas others use separate lines for read and write data so that data is always transferred in only one direction on any set of lines. Others dispense with the tristate drivers altogether, favouring a gate-multiplexed datapath which will typically operate much faster than a tristated solution. Such schemes don't share any wires in the sense usually associated with a shared bus, but their protocol is usually very similar.

Furthermore, whichever datapath approach is taken, the timing of the data validity varies between systems. Some pass the write data at the same time as the address; others pipeline the transfer so that read or write data is always passed after the address.

3.3.4 Arbitration

Shared buses only have one data highway to provide transport between many nodes.
To avoid corruption (and in some technologies also to avoid damaging short-circuits),
it is mandatory that only one device drive the bus at once. Arbitration is the technique
of ensuring mutually exclusive access to the bus. Again there are a number of
approaches:

- Collision detection - where a bus client has to check for corruption of any data
  it sends onto the bus, and if corruption occurs (due to a collision with another
  client also using the bus at the same time) then the client retries the transfer
  after some delay. Such schemes do not enforce mutually exclusive use of the
  bus, but do ensure that data is passed correctly. 10-base-2 Ethernet uses this
  scheme. This technique is satisfactory for low occupancy buses in technologies
  where collisions (and multiple drivers active at the same time) can be tolerated.
  However, as bus occupancy increases, so does the number of collisions, and
  performance (in terms of successful transfers per second) is degraded.

- Distributed arbitration - whilst collision detection is in effect a distributed form
  of ensuring that data is transferred uncorrupted, distributed arbitration is
  usually taken to mean a system of negotiating who owns a shared resource
  where the negotiation circuits are distributed between the contenders. This can
  be clarified by an example: a (narrow) SCSI bus can have one of eight possible
  owners at a given time. When a device requires the bus, it places its unique
  SCSI-ID on the bus during the arbitration phase of the protocol (which lasts for
  a fixed duration). If more than one initiator contends for the bus, then the one
  with the highest unique-id wins.

- Centralised arbitration - gathers the choice-making logic into one unit, which
  communicates with each possible owner of the shared resource using a
  dedicated connection that uses a request/grant protocol (usually very similar to
  the 4-phase handshakes found in asynchronous design).

When dealing with the allocation of a critical shared resource, the issues of
fairness, priority and starvation, as discussed in any good operating-systems design
text, are all relevant. They can be accommodated to some degree through the design
of the arbitration protocol for a distributed scheme or the arbiter design in a centralised
approach.

3.3.5 Atomic sequences

Some actions require that an initiator is allowed to make a number of consecutive
transfers across the bus or to a specific target without other devices using that resource
in the meantime. These transfers are said to be *atomic* or *locked*. One use of such
transfers is in implementing the semaphore operations used by operating system
software.

3.3.6 Bursts

Similar to the concept of an atomic transfer sequence is the burst transfer. Here, a number of related transfers are performed at an accelerated rate, often by transmitting one address packet and then performing a number of data transfers. Such bursts are often used for cache line refills where a known number of data-words must be transferred from the same page of memory. Burst transfers usually allow a higher throughput than when performing the same operation using multiple distinct transfers. The higher performance results from reducing the number of arbitrations (and possibly address decodes) that have to be performed. However, because a burst occupies the bus for multiple transfers, it can have an adverse effect on the latency encountered by other initiators awaiting access to the bus.

3.3.7 Interlocked or decoupled transfers

Buses where the command/address phase and the response/data phase of the transfer are tightly coupled are known as interlocked buses. This approach usually leads to simple synchronous control circuits. The alternative approach is to provide minimal coupling between the two phases of a transfer, allowing the command/address and response/data to be passed separately. A separate arbitration stage is then required for each part of the transfer.

3.3.8 Split transactions

To improve bus availability, slow devices may be allowed to accept the address/ command part of a transaction and then disconnect from the bus, e.g. SCSI. They then reconnect later and perform the data action of the transaction. In the meantime, the bus is available for transfers between other devices. The transaction is said to be a *split-transaction*. Such transactions can be implemented on top of either an interlocked or a decoupled protocol and may require two transfers per transaction, one to pass the command/address and one to return the read-data and any other response bits. A split transaction can be mapped directly onto a single decoupled transfer in some cases.

3.4 Interconnect performance objectives

The key performance requirements of an interconnect solution are:
- Low latency - essential whenever there is a dependency on the transfer. A data load by a CPU is a good example, since the processor may be stalled until the data is returned (although in AMULET3 the reorder buffer allows the processor not to stall until a data dependency arises). The latency of fetching instructions is important when a change of instruction stream is encountered, due to a (non-predicted) branch or interrupt but has a lesser effect when

executing sequential code due to the speculative prefetching that is now commonplace.

- High throughput - necessary when large quantities of data are to be transferred, such as when a cache line is being reloaded from main memory or for the instruction fetch process of a microprocessor.

These two goals tend to conflict, a common example being the addition of extra pipeline stages to a system which often increases the throughput at the expense of the overall pipeline latency.

3.5 Commercial on-chip buses

Over the past two years, a large number of on-chip buses have been proposed by the many companies active in the SoC design industry. Three of the more notable are the Peripheral Interconnect (PI) Bus, AMBA and CoreConnect. A short precis of these three buses are provided in the following sections.

3.5.1 Peripheral Interconnect Bus (PI-Bus)

The OMI PI-Bus [64] is an on-chip bus developed in a (European) project to standardise on-chip interconnect. Like most demultiplexed buses, the PI-Bus consists of an address bus and a data bus. Each bus is scalable up to 32 bits in size, and the specification allows for a clock frequency of up to 50MHz (on 1996 technology), allowing a maximum throughput of 200MB/s. This is aided by the pipelining of transfers which allows the address phase of a transfer to occur at the same time as the data phase of the previous transfer.

PI-Bus supports the usual range of bus features such as a multimaster capability, atomic transactions, transfer deferral and data accesses of either 8, 16 or 32 bits. Furthermore, the protocol includes the necessary acknowledge codes to allow a split transaction to be constructed using multiple bus transfers, and there is provision for flexible burst transfers of an arbitrary length.

3.5.2 Advanced Microcontroller Bus Architecture (AMBA)

AMBA [2,3,38] is a collection of on-chip buses from ARM Ltd for satisfying a range of differing criteria:

- AMBA-APB - The AMBA Advanced Peripheral Bus is a simple strobed-access bus with minimal interface complexity. It is suitable for hosting many peripheral functions (with only one initiator - often a bridge to one of the other members of the AMBA family), and holds the address and control signals valid throughout a transfer, not supporting any pipelining.
- AMBA-ASB - The AMBA Advanced System Bus is a multimaster synchronous system bus. It operates with a pipelined protocol such that arbitration for the next transfer can occur in parallel with the current address

transfer, which in turn will be occurring at the same time as the previous data transfer. The AMBA-ASB supports either a central multiplexer interconnection scheme, or a tristated approach. The AMBA-ASB protocol is very similar to that of the OMI PI-Bus discussed above.

- AMBA-AHB - The AMBA Advanced High-performance Bus is a high throughput synchronous system backbone that was added to AMBA in mid-1999 with the release of version 2.0 of the specification [3]. It supports advanced features including burst transfers and split transactions (with a maximum of 16 initiator units, each of which can have one outstanding command) with separate datapaths for read data and write data. It uses single clock edge operation with a multiplexed data path (instead of tristated bus lines) to provide for high frequency operation and simplify the automated synthesis of AMBA-AHB systems. To further boost performance the address decoding is performed by a centralised unit and wide datapath configurations are available, up to 128 bits.

A typical AMBA system will contain either an AHB or an ASB bus hosting the microprocessor, a DMA controller and on-chip ROM/RAM, with an APB bus used for the connection of simpler peripheral devices.

Both the ASB and AHB are multimaster system buses and thus support transfer deferral for the avoidance of deadlock situations (as when bridging between two buses, and described further in Section 5.6.4 where the handling of this in an asynchronous environment is addressed). The ASB allowed the use of this same mechanism for the implementation of split transactions, albeit in a crude manner, whereas one of the major features of the AHB is its direct support for split transactions allowing them to be performed much more efficiently. This is possible because the bus-arbiter has been given more intelligence and a direct connection to each target device, allowing the target to indicate when an initiator should be allowed to retry its transfer (and hence be granted the bus again), rather than requiring the initiator to poll the target using the bus as is the case for AMBA-ASB and PI-Bus.

3.5.3 CoreConnect

The final on-chip bus reviewed here is CoreConnect [39,44,45] from IBM. As was the case for AMBA, CoreConnect is a family of three (synchronous) buses designed to meet different performance criteria:

- CoreConnect DCR - To provide for configuration register access, the CoreConnect architecture defines a separate low bandwidth Device Control Register bus connecting all of the devices on the PLB, including the PLB-arbiter and the PLB-OPB bridge. This allows slow/infrequent configuration operations to proceed without impeding activity on the high throughput PLB.

- CoreConnect OPB - The CoreConnect On-Chip Peripheral Bus protocol is similar to that of both the OMI PI-Bus and AMBA-ASB including the usual 32-bit address and data buses, overlapped arbitration, and bus parking so that one initiator can be given a default ownership of the bus when it is idle so as to reduce that initiator's latency on a subsequent access.

- CoreConnect PLB - The CoreConnect Processor Local Bus is the high-performance member of the family. It uses pipelined/overlapped arbitration, address transfers, read data transfers and write data transfers (using separate pathways for read and write data) combined with burst transfers to perform a maximum of two data transfers per clock cycle (if both a write and a read are active at once). Flexible DMA support allows either store-and-forward (where the DMA controller first reads and then writes the data in two separate transfers) or flyby (where the DMA controller issues the addresses for both a read and a write, and then the data is transferred directly between the target devices) DMA techniques to be used. The PLB supports bus-widths up to 128 bits.

None of the CoreConnect family use tristate bus lines, using the alternative gate multiplexing approach to pass the appropriate signals onto the bus. The different buses in the CoreConnect family are allowed to operate at different frequencies, but all of the clocks must be derived from a common source so that a common clock edge can be used when bridging between the buses.

3.6 Summary

The basic principles and issues involved in the connection of multiple system components have been introduced, and solutions including the shared bus have been reviewed. However, for whatever reason (possibly some of the ones in Section 1.2), all current commercial on-chip buses are synchronous and do not address the needs of a mixed synchrony chip which may include asynchronous components. The remainder of this book investigates the design of an asynchronous System-on-Chip bus and its benefits relative to a synchronous bus for the interconnection of asynchronous macrocells. As an example of the viability of asynchronous on-chip buses, MARBLE (a 2-channel asynchronous on-chip bus) is presented in the context of the AMULET3H asynchronous subsystem in Chapter 8.

4.   The Physical (Wire) Layer

A shared bus is a collection of wires where all interfaces to the wires comply with an ordered protocol devised to avoid deadlocks and data corruption. In the layered bus implementation hierarchy of Figure 4.1 these wires collectively form the lowest layer, the physical layer.

The physical layer of this bus hierarchy defines how the wires will be terminated, their separation and their size. These issues are addressed for an asynchronous SoC bus in this chapter.

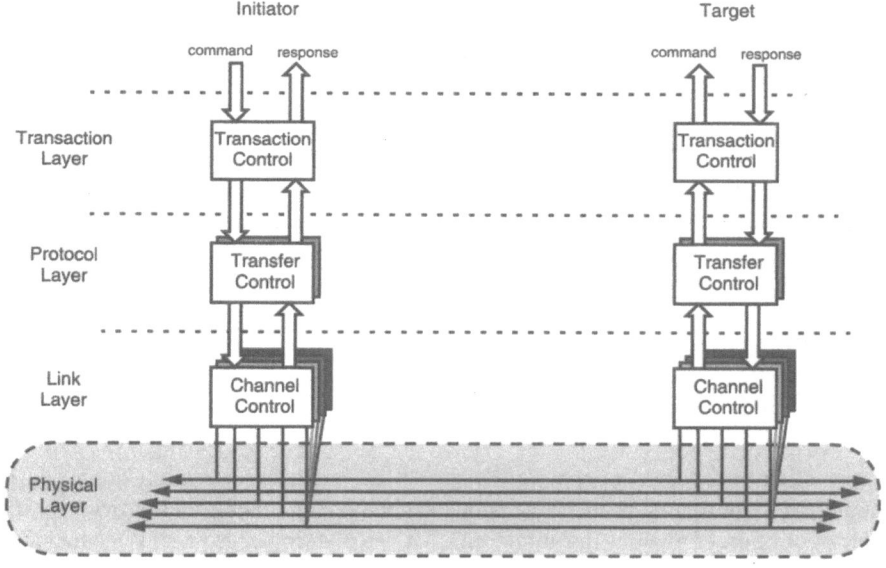

Figure 4.1: A layered bus hierarchy

4.1 Wire theory

Although a wire in a circuit is often considered to have negligible resistance and capacitance, these assumptions are no longer true with submicron silicon feature sizes and all of the parameters of the wire must be considered. Since the circuit elements of a wire are distributed along its length the description of the wire requires a partial differential equation. Consider the model of a wire of length δx shown in Figure 4.2 (with resistance, capacitance, inductance and conductance per unit length of R, C, L and G respectively). The conductance G is negligible in silicon chips since SiO_2 is an excellent insulator. The behaviour of the voltage, V, on the wire as a function of position (x) and time (t) is then given by Equation 4.1 [25].

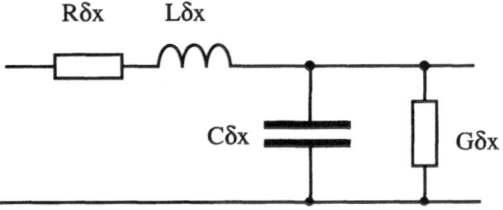

Rδx   Lδx

Cδx   Gδx

Figure 4.2: Infinitesimal length wire model

$$\frac{\partial^2 V}{\partial x^2} = RC\frac{\partial V}{\partial t} + LC\frac{\partial^2 V}{\partial t^2}$$
Equation 4.1

This equation describes the propagation of a signal along the wire by two mechanisms, diffusion and travelling waves. For on-chip wires the effect of L is very small compared to that of the resistance, and the travelling wave term is thus insignificant [25]. Signal propagation can therefore be considered to be by diffusion along an RC transmission line. The delay of a signal along the line is then quadratic with the line length (because both resistance and capacitance are proportional to the length of the line) and signal edges are widely dispersed on long lines.

4.2 Electrical and physical characteristics

The electrical properties of a wire that affect the signal propagation described by Equation 4.1 are related to the dimensions of the wire. As mentioned previously, the inductance of on-chip wires in submicron CMOS technologies is generally considered negligible. However, resistance and capacitance are both significant, especially for long interconnects.

The MARBLE bus architecture presented in this book has been implemented on a 0.35 micron CMOS silicon process (known as VCMN4) from VLSI Technology, Inc. which allows up to three metal layers (known here in the order of increasing separation from the substrate as metal-1, metal-2 and metal-3). Consider the low-level characteristics of this process and their implications for the design of the interconnect.

The important choices to be made here are the:

- wire width;
- wire separation;
- wiring layers to use.

To answer these questions requires a consideration of the resistance and capacitance of the wires, and how these affect both the signal delay and the crosstalk between signals on different wires.

Figure 4.3 shows the resistance (in ohms) and the lateral and interlayer capacitances (in fF) per mm length of a group of 0.7μm wide (minimum width) conductors separated laterally by 1.1μm (the minimum allowable for the metal-3 layer) in the VCMN4 silicon process. Tracks on the metal-2 and metal-3 layers are vertically aligned with those on the metal-1 layer. The values shown were obtained by post-layout extraction using the Compass Design Tools [20].

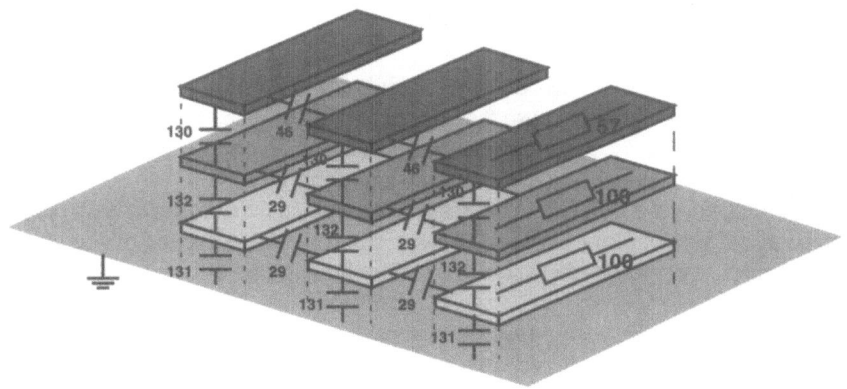

Figure 4.3: Resistance and capacitance for nine 1mm long close-packed wires

4.3 Termination

The diffusive nature of transmission along an on-chip line means that the issues of termination and reflection [57] (as must be addressed for off-chip buses which use travelling wave propagation as their principle mode of signal transfer) can be avoided. The only consideration is that all floating lines should be driven by weak feedback keepers to avoid the power loss that would result from intermediate voltages causing short-circuit currents in bus voltage sensing logic at each of the bus interfaces.

4.4 Crosstalk

Electrical charges affect their environment both through their presence and through their motion. The effects of a moving charge on neighbouring wires in a circuit occur due to inductive coupling and capacitive coupling. These effects are typically weak, but as CMOS process technology moves to smaller feature sizes the closer proximity of the wires and transistors means that they become more significant (especially the capacitive coupling). Coupling can give rise to effects such as:

- spikes or glitches on a signal due to activity on a neighbouring signal;
- slowdown or speed-up of signal edges due to activity on the surrounding signals.

These effects, commonly called crosstalk, cause serious problems for both synchronous and asynchronous designers. In both cases, slower edges mean that both the n-stack and the p-stack of CMOS gates can be conducting for longer, leading to higher power consumption. Crosstalk also affects the performance of the circuits:

- in synchronous design, the slowdown of signal edges affects the maximum usable clock frequency;
- in asynchronous design the matched-delay path, used to ensure that the request of a bundled channel arrives at the receiver no earlier than the data, must allow for crosstalk effects.

The spikes or glitches caused on a wire by crosstalk are a form of noise. Depending on where they happen they can cause the false triggering of gates or latches, upset state machines, and possibly lead to system failure (e.g. the glitch may be interpreted incorrectly as a handshake event on a request signalling line).

Crosstalk effects are difficult to determine through static timing analysis because they depend on the coupling between the wires and the edge-speeds of the signals. Much has been written on the subject of modelling these effects, and the IEEE Computer Society sponsors an annual workshop on *Signal Propagation on Interconnects*. Theory indicates that designers should be careful to keep adjacent wires short and well-separated wherever possible, since the coupling between the wires is proportional to their overlap length. However, for long bus lines the designer needs to know how close the wires can be packed, since such lines are still commonly routed manually to minimise their area requirement.

Figure 4.4 illustrates the severity of the crosstalk problem in an extreme (but not unimaginable) case. This shows the amplitude at the far end of a 10mm metal-2 line surrounded on all three layers (above, below and the same layer) by other parallel conductors as illustrated in Figure 4.3, each line driven by a separate inverter.

The three panes in Figure 4.4 show:

- In the upper pane: the output from the far end of the wire. Crosstalk causes the large overshoots and different edge speeds for different sets of transitions in this system.
- In the middle pane: the signals at the node between the drive inverter and the wire.
- In the lower pane: the input to the wires' drive inverters (observe the clean edges).

The plot in Figure 4.4 and the other results in this chapter were all generated using SPICE [42] simulations for the 0.35 micron VCMN4 process technology [86] used for the MARBLE bus and AMULET3H chip described in Chapter 8. The model of the channel used in the simulations was constructed from 0.5mm segments similar to those shown in figure 4.3. These results are corroborated by the theoretical values derived elsewhere [61] from the solution of Maxwell's equations.

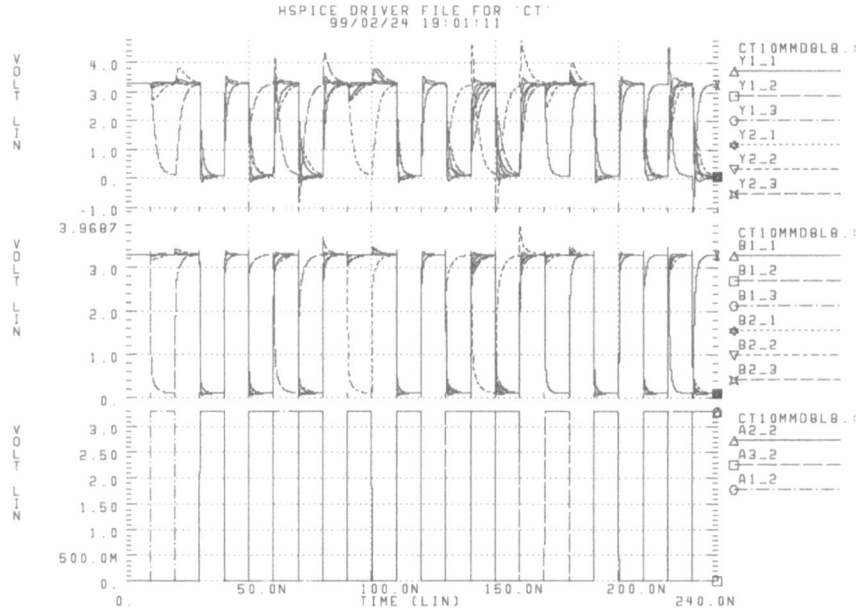

Figure 4.4: SPICE plot of 9-wire system

4.4.1 Propagation delay for well separated wires

The graph in Figure 4.5 shows the delay from the input of the drive inverter crossing the half-rail voltage to the far end of a single wire on the metal-1 layer (closest to the substrate, hence highest loaded) with no other surrounding wires (hence no crosstalk effects) crossing the half-rail voltage. There are four visible groups of curves which, coming down the page, correspond to drive inverters of strengths 2x, 4x, 8x and 16x the strength of a minimum sized inverter in this technology. Within each group there are four traces which, moving up the page, correspond to gate-loading of the wires (lumped at the far end) of 2x, 4x, 8x and 16x the load presented by a minimum width inverter in this technology. There are no curves corresponding to the use of a minimum width drive inverter as its drive strength is only sufficient to drive the inputs of two or three small gates with minimal wiring load.

This graph confirms that with small gate loads (up to 16 times the load presented by a minimum width inverter) and long wires:

- the load (and hence signal delay) due to a few gates is small compared to the load due to the wire capacitance;

- a driver 16 times the size of a minimum sized inverter should be sufficient to give reasonably small delays.

Thus, for the remainder of this analysis, all simulations use drivers and loads of 16 times the size of a minimum sized inverter in this technology (this was also the largest tristate driver available in the standard cell library used for this work).

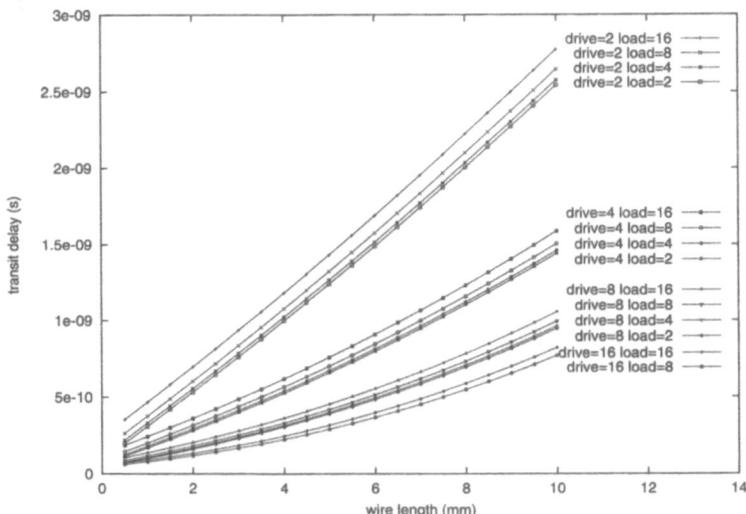

Figure 4.5: Signal propagation delay (transit time) for an isolated wire

4.4.2 Signal propagation delay with close-packed wires

The most densely packed formation for running bus wires around a chip is to use all three metal layers, with the minimum allowed spacing between wires as illustrated earlier in Figure 4.3. A test system with nine wires packed in such an arrangement was simulated. The graph in Figure 4.6 shows the delay (including the driving inverter) from the input of the drive inverter to the far end of the wire for the central wire of the metal-2 layer (i.e. the wire at the centre of the 9-wire bundle). The delays were measured between the signals crossing the half-rail voltage. This graph when compared to the equivalent "drive=16, load=16" curve in Figure 4.5 (for the same wire and drivers in isolation) shows that:

- the delay of a surrounded wire is about double that for an isolated wire, as expected since the capacitive loading has slightly more than doubled (since most of the capacitive coupling is interlayer, not lateral in this technology);
- a significant variation in delay, of up to 2ns is possible as a result of crosstalk.

Thus even if the request line of a channel were well separated from other wires, an additional delay of around 1ns would still be required to meet the bundling constraint.

4.4.3 Alternative wiring arrangements

Although wires may be routed this densely for long point-to-point connections, the situation could be improved by adding additional amplification at intermediate points along the path, maybe every 2mm or 4mm. However, with a multipoint bus where the wire can have different drivers at different points in time, this is not possible. Also,

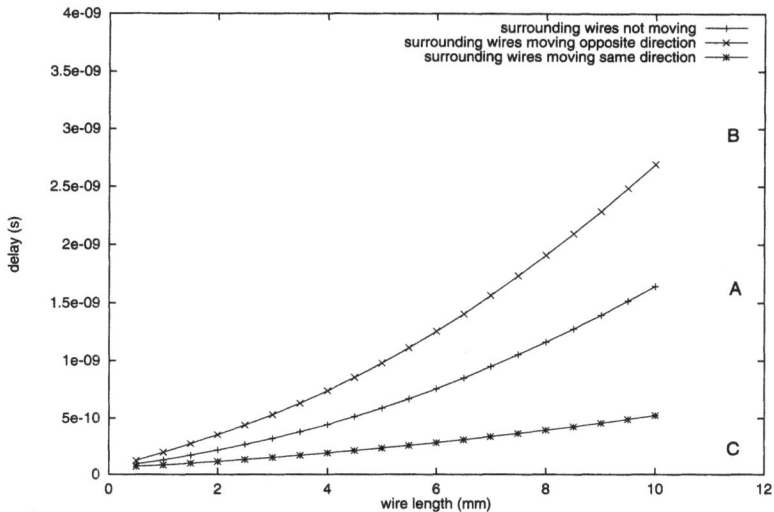

Figure 4.6: Signal propagation delay (transit time) for a close-packed wire

because of the need to tap onto the lines at multiple points, the wires cannot be routed so closely in all three layers (since the tap-off connection requires a metal layer for its wires). This raises the question of "what wire formation and separation should be used for long wires such as are used in MARBLE?". To answer this question for the AMULET3H local buses and MARBLE, a variety of formations and wire separations as illustrated in Figure 4.7 were simulated to investigate the effects of:

- increasing the spacing between the wires;
- not using all three layers;
- staggering the wires between layers, so that wires on one layer are not directly above or below wires on an adjacent layer.

Figure 4.7 also shows the results of these simulations by plotting the delay (range 0 to 3ns) versus wire length (range 0 to 10mm), each graph showing three curves:

- in the upper curve: the worst case delay encountered when a signal is changing level in the opposite direction to the transitions being made by its surrounding signals;
- in the middle curve: the normal delay encountered when a signal changes level and its surrounding signals remain unchanged;
- in the lower curve: the best case delay encountered when all of the signals are changing level in the same direction.

Crosstalk is responsible for the variation in delay between the two extremes of the uppermost and lowermost curves on each graph.

The labels in Figure 4.7 are formed according to Table 4.1 to show which wires were present in which arrangement, their separation and width, and if the position of the wires on each layer was staggered or the wires were vertically above each other.

Table 4.1: Legend codes for Figures 4.7 and 4.8

Legend code	Meaning
mx	3 parallel wires on the metal-x layer (layer 1 nearest to substrate, layer 3 furthest from substrate)
sbc=xxx	spacing between centres of the wires in μm (minimum of 1.6)
w=xxx	wire width in multiples of 0.7μm (the minimum wire width)
st	adjacent layers were staggered by half the separation so that wires were not vertically aligned

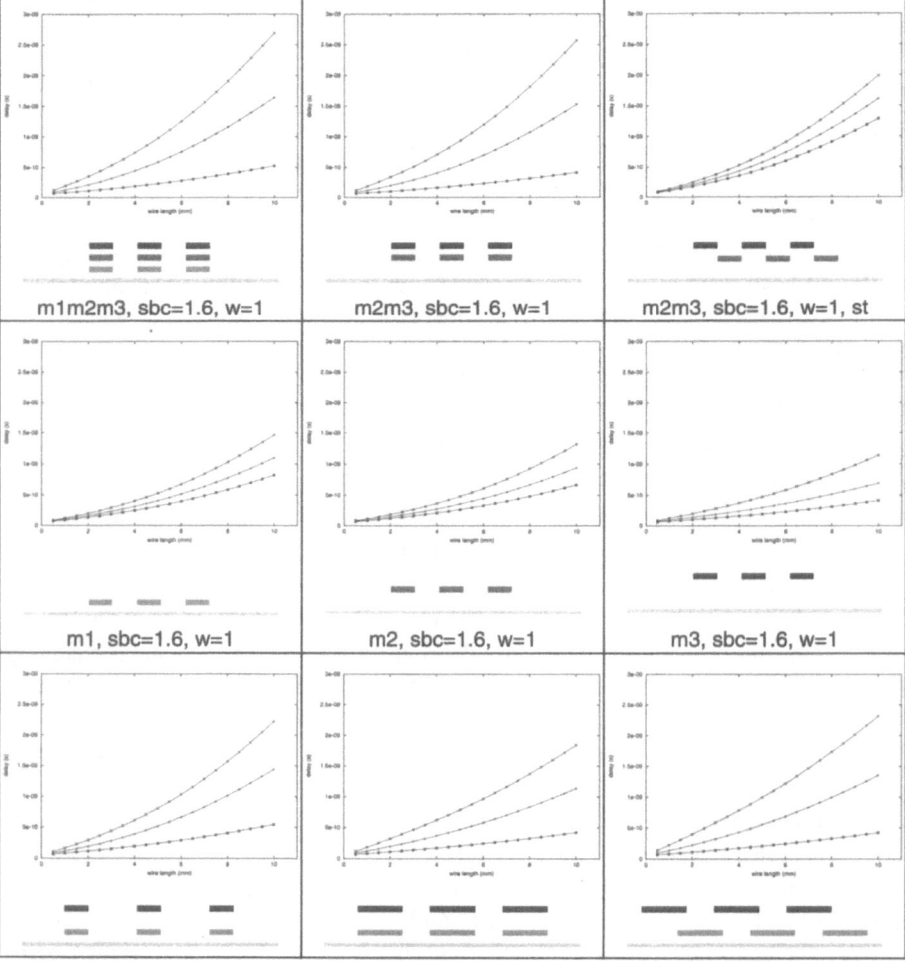

Figure 4.7: The effect of wire formation and crosstalk on transmission delay

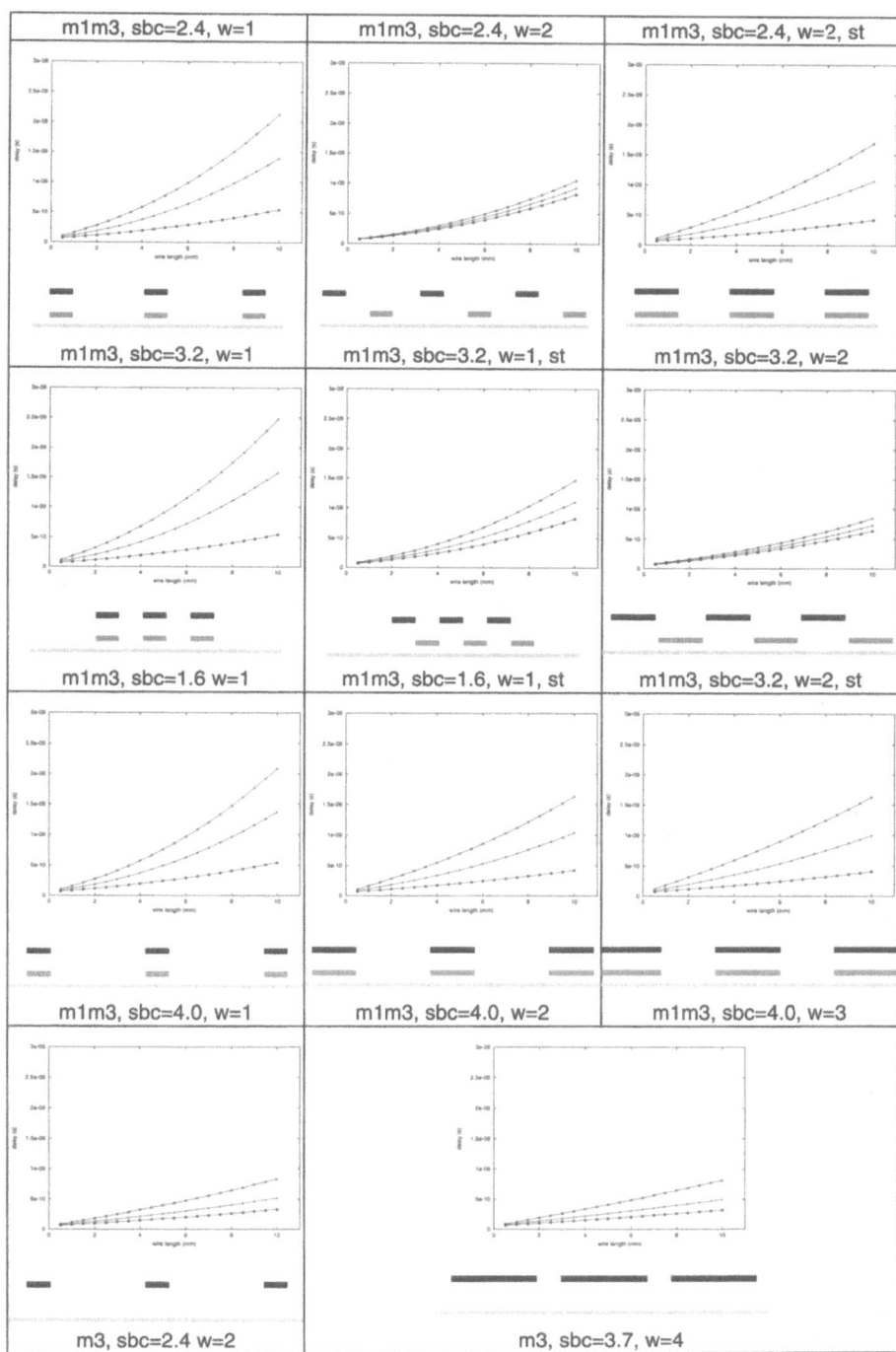

Figure 4.7: The effect of wire formation and crosstalk on transmission delay

One could make a choice of wiring formation purely on the grounds of the basic delay and the delay variation as shown in the above graphs, but usually silicon area is at a premium. To allow performance to be traded against area usage, the worst-case total delay along a 10mm wire was plotted against the width of the silicon area occupied by the wire (averaged over 36 wires - allowing for a 32-bit data bus and some signalling wires). The resulting graph is shown in Figure 4.8.

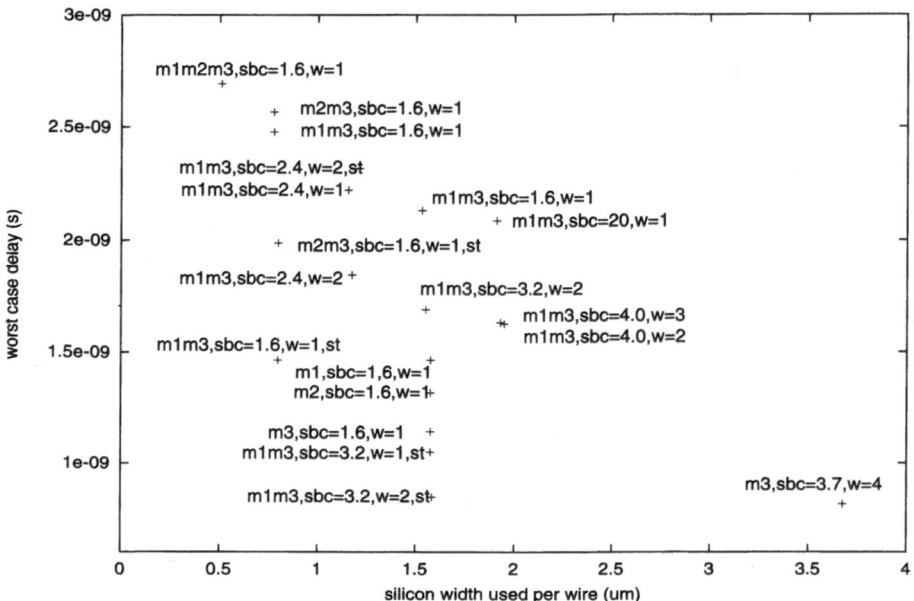

Figure 4.8: Delay versus datapath width per wire

4.5 Summary

Signal transmission over a set of wires in submicron CMOS technology must be considered in the design of long interconnects since the resistance and capacitance of the wires can cause considerable transmission delays and crosstalk effects.

The simulations presented in this chapter show that for maximum speed, the result is as one would expect: use the metal-3 layer with wide (2x minimum, i.e. 1.4μm wide should be sufficient) wires separated by at least 2.4μm between centres. This will give a delay of about (0.5±0.4)ns for a 10mm length with a separation of 2.4μm between wire centres. This configuration was used for the processor local buses in the AMULET3H chip.

However, the best wire-density vs performance trade-off is obtained by using both the metal-1 and metal-3 layers, with double the minimum width tracks, double the minimum spacing between centres and staggered centres to achieve a delay of (0.75±0.15)ns for a 10mm wire. The wires in the MARBLE bus are actually around 5mm long, for which this configuration gives a delay of (0.35±0.05)ns.

5. The Link Layer

Most asynchronous VLSI channels, as introduced in Chapter 2, are designed to cover short distances and provide a point-to-point connection between one sender and one receiver, with data transferred in only one direction.

The design of an asynchronous macrocell bus requires a new type of channel, one that can connect many ports, supporting transfers in either or both directions. This chapter illustrates the problems encountered in the design of such a channel, and gives practical solutions to those problems. The multipoint channel introduced here forms the link layer of the bus hierarchy shown in Figure 5.1.

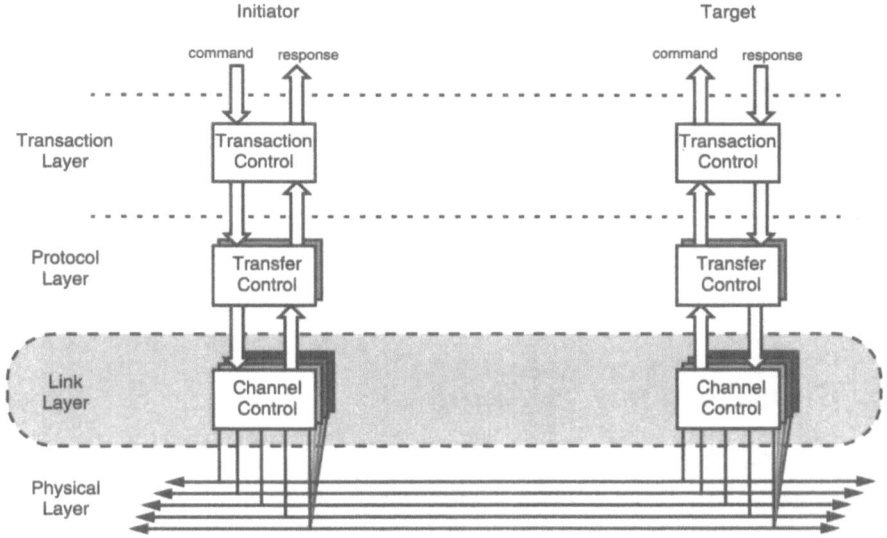

Figure 5.1: A layered bus interface

5.1 Centralised vs distributed interfaces

The interfaces to a shared bus may be centralised or distributed, with consequential effects on the size and performance of the system and the design of the interfaces themselves.

- The *centralised* approach groups all the bus interfaces and control components in a central hub as shown in Figure 5.2a, forming a star network. Each device has dedicated point-to-point connections to the hub. This approach is more expensive in area but minimises the length of the shared lines, and hence their load, allowing faster edges and weaker drivers.

• The *distributed* approach, as shown in Figure 5.2b, represents the more conventional view of a shared bus. This approach places the bus interfaces near to the devices so as to minimise the size of the point-to-point links between the device and its interface. It leads to a higher loading on the bus and slower operation, but typically gives a much smaller implementation due to the reduction in wiring.

In both cases, the same issues must be addressed as discussed below.

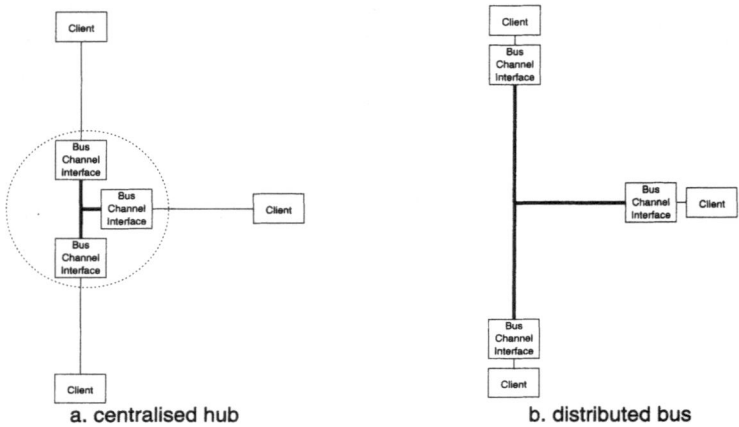

a. centralised hub   b. distributed bus

Figure 5.2: Centralised and distributed interfaces

5.2 Signalling convention

The issue of 2-phase versus 4-phase signalling has been addressed in the context of conventional point-to-point unidirectional channels, as summarised in Chapter 2. However, the significantly different nature of a multipoint channel warrants the reconsideration of the issues involved in the choice of signalling protocol.

Whilst 2-phase design initially appears to be a good idea for a multipoint channel, since it minimises the number of signalling edges and hence also the cycle time and power dissipation, the more difficult to detect quiescent state complicates bus hand-over and turn-around. This is because the signalling protocol affects the state of the channel after a transfer in that:

• 2-phase signalling causes the signal lines to toggle state after each cycle;

• 4-phase signalling leaves the signal lines in the same state as before the cycle.

As a result, 2-phase signalling suffers relative to 4-phase signalling from an increased complexity of the logic within the channel (the MERGE elements in Figure 5.3) used to generate the channel request and acknowledge signals that are distributed on the channel from the individual signalling outputs of each device. Further details are presented in Sections 5.6.2 and 5.7.1.

Data drive hand-over is similarly affected by the choice of signalling protocol. Here, there is a requirement to avoid drive clashes between one device switching off its drivers and the next device switching its on. With the 2-phase protocol there are insufficient signalling edges to define an *undriven* phase for the data wires, whereas with the 4-phase channel this can easily be achieved. Further details are contained in Sections 5.6.3 and 5.7.4.

5.3 Data encoding

The choice of single-rail or dual-rail presents a similar problem, in that again there are advantages in both styles. Currently, however, for wide parallel interconnects the costs of a dual-rail system (twice the wiring area and additional logic to detect the presence of a valid word on the channel) outweigh its benefits. Some of the issues involved in designing a multipoint dual-rail channel using handshake circuits [81] were addressed by Molina [58,59] but the overhead of using a dual-rail approach is too large for use with connections such as those of a macrocell bus.

For these reasons, this chapter addresses the design of a 4-phase single-rail multipoint channel.

5.4 Handshake sources

For a conventional unidirectional point-to-point channel which device signals events on the request wire and which issues the acknowledge event is unimportant. One combination requires a push protocol, the other a pull protocol, but communications are always between the same two devices and in the same direction.

A multipoint channel, such as that shown in Figure 5.3, connecting three devices A, B and C presents two complications:
- the initiator and target can be different for each transfer;
- the direction of each data transfer is not known before it occurs.

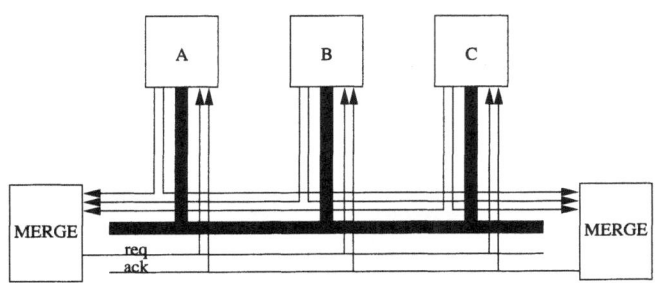

Figure 5.3: Multipoint (bus) channel wiring

There are thus two practical alternative schemes for controlling which device will deliver which events:

- the sender provides request events, the receiver provides acknowledge events;
- the initiator provides request events, the target provides acknowledge events.

The first option allows data always to be pushed by the sender, but requires every device on the channel to be capable of delivering request or acknowledge events depending on which direction data is being transferred. This leads to complex circuits for determining which unit will provide each signalling event.

The second option gives simpler circuits due to the isolation of the request and acknowledge control functions, even though it requires that the channel allows the initiator to push data to the target or pull data from the target depending on the transfer direction.

5.5 Bidirectional data transfer

Multipoint (bus) channel actions often require both push and pull actions to be performed in a single communication allowing simultaneous information transfer in two directions. Typical pushed information includes:

- an originating initiator identifier;
- an address/target identifier;
- the transfer size;
- a transfer action/direction indicator;
- the data payload.

Typical pulled information may include:

- the data payload;
- status bits.

Simultaneous push and pull communications between two devices can be implemented using two conventional push channels. However this requires four signalling wires to implement the bidirectional information exchange. By merging the protocols of a push and a pull channel, as illustrated in Figure 5.4, a bidirectional exchange can be performed in one cycle on a single channel using only two signalling wires, with separate wires for the push and pull datapaths.

This is a worthwhile saving in wiring for long channels or multipoint channels with many initiators and/or targets, and simplifies the channel control functions for the bus channels.

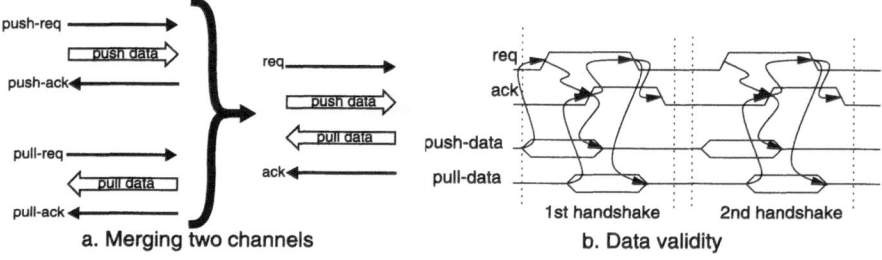

Figure 5.4: 4-phase channel with both push and pull data paths

To avoid data clashes, separate data wires must be used for the push-data and pull-data. Alternatively, the wires may perform one function in one cycle and the other function in a different cycle.

5.6 Multiple initiators on one channel

When a channel has more than one initiator, extra control complexity is required to determine which initiator should use the channel and to ensure a clean hand-over of ownership of the channel from one initiator to another. This is the role performed by the bus arbitration logic.

5.6.1 Arbitration

For correct operation of the channel avoiding drive clashes, data-corruption and signalling failure, it is imperative that only one initiator acts upon the channel at any one time. Arbitration is thus required between initiators to determine which *owns* the channel. The distributed arbitration techniques found in off-chip buses or networks such as SCSI [77], Ethernet or the Trimosbus [76] are not suitable for low-power CMOS on-chip systems because they permit drive-clashes or polling of the arbitration signals. Consequently synchronous on-chip buses (e.g. AMBA [2]) and some off-chip buses (such as PCI [72]) use a centralised arbitration system with request and grant handshaking signals (typically using a 4-phase protocol) connecting the device to the central arbiter. This argument also justifies the use of a centralised arbitration system in an asynchronous CMOS VLSI bus. Such a system can be built around instances of the *mutex* (mutual exclusion) structure as proposed by Seitz [71] and shown earlier in Section 2.2.7.

In-line arbitration
The multiple initiator function of the channel is logically equivalent to an arbitrated call block (Figure 5.5 shows a 2-input arbitrated call). This formation provides mutually exclusive access to the output channel (req0/ack0) for the two input channels (req1/ack1 and req2/ack2), ensuring that one complete cycle is performed on the output channel for each input channel cycle.

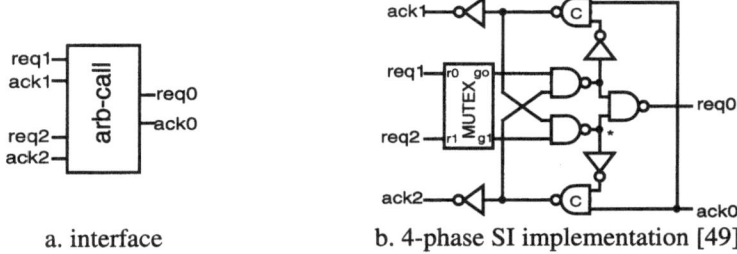

a. interface   b. 4-phase SI implementation [49]

Figure 5.5: Arbitrated call

In the multipoint bus channel, there is also data that has to be steered through the call block from the input channel to the output channel, a feature which can be added to the arbitrated-call block by connecting the control input of a multiplexer to the output of the NAND gate marked * in Figure 5.5b.

Pipelined arbitration

The arbitrated-call approach to allowing multiple initiators to access the central bus channel has a significant disadvantage: arbitration and bus accesses are sequential; consequently the bus is idle whilst arbitration for the next cycle occurs, thus affecting the overall throughput of the channel.

In a high performance system it is possible to hide the latency of the arbiter by separating the arbitration function from the call and multiplexing functions. Arbitration for the next channel access can then proceed in parallel with the current cycle. The grant signal is thus an *early-grant*, indicating which device will take ownership of the channel when it next becomes idle, as shown for one initiator by the STG in Figure 5.6.[1]

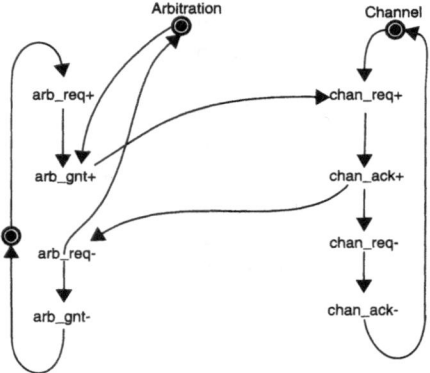

Figure 5.6: Hidden arbitration

This means that the arbitration network can be optimised to give a fast grant to the first contender for an empty bus, since all subsequent arbitrations, whilst the bus is in use, are hidden by the current transfer and do not affect either latency or throughput provided they have occurred by the time the bus is available.

---

1. The arc from chan_ack+ to arb_req- could have been from chan_req+ to arb_req-. Observing the acknowledge event rather than the request event here includes the delay for the event to propagate down the channel request wire, through the target and back up the acknowledge wire. This is assumed to be sufficient delay to allow all gates connected to the chan_req wire to observe and process the chan_req+ event before a subsequent chan_req- event occurs. This technique is used in a number of instances in the design of the multipoint channel.

N-way arbiter
Current VLSI technology allows the implementation of only 2- or at most 3-way [51] mutexes, although the robustness of the latter is not proven. These can be used directly with buses with the corresponding number of initiators to provide very low latency arbitration. Arbitration between a larger number of contenders can be achieved with a network of multiple mutexes connected using the approaches described below.

Mutex cascade
A larger mutex can be constructed from a cascade of 2-input mutexes using a multistage approach. Figure 5.7 illustrates how 3- and 4-way arbitration can be implemented using mutexes alone. To generate a grant, a request has to arbitrate with all other contending requests (or their delayed version after passing through a mutex). Thus for 3-way arbitration, a request has to pass through two mutexes, and through three mutexes for 4-way arbitration. This scheme gives low latency, but the hardware requirements grow quadratically with the number of contending input requests.

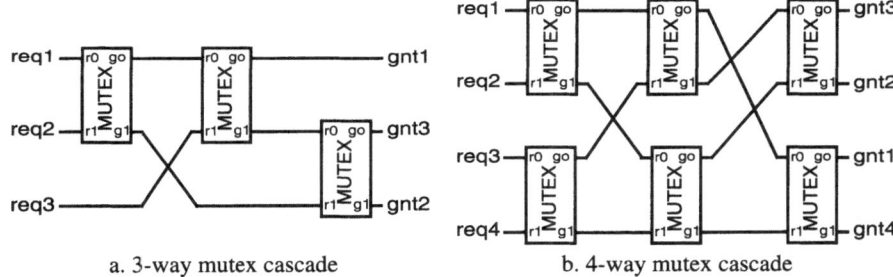

a. 3-way mutex cascade   b. 4-way mutex cascade

Figure 5.7: Arbitration using cascaded mutexes

Token ring arbiter
A form of arbitration well suited to a distributed approach is the token-ring arbiter [54,55]. The circuit for a ring-arbiter element is shown in Figure 5.8a, and a four-way ring arbiter constructed from such elements is illustrated in Figure 5.8b. One of the elements in the ring is initialised with its latch set (giving it the token), all others having their latches reset.

When an element's input request, req, is asserted, the state of its latch is checked. If the token is held (latch set) then a grant is given on gnt, if not then the token is obtained by performing a cycle on the rreq/rgnt channel. The S-element [81] encloses the complete output cycle in the first part of the arbitrated-call output handshake. The lreq/lgnt channel allows another ring-arbiter element to ask for the token from this element, the arbitrated call resolving conflict between requests on the lreq and req inputs.

However, because the token in such systems has to be passed around the ring on demand, the worst case arbitration latency grows linearly with the number of possible contending devices, as does the hardware required for implementation of the ring.

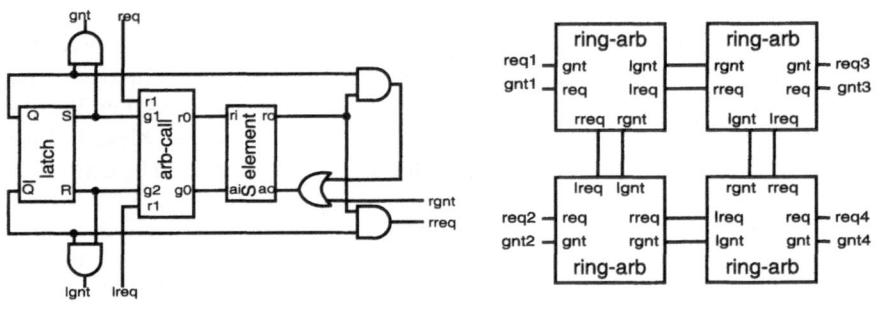

a. Ring arbiter element (after [55])   b. 4-way ring arbiter

Figure 5.8: Ring-arbiter

Tree arbiter

A tree arbiter is composed of elements similar to an arbitrated call but uses eager request propagation [49] to allow the arbitration at each level of the tree to be performed in parallel with activity at higher levels, instead of in series as when a tree is formed from a set of standard arbitrated calls.

Josephs proposed a speed-independent custom tree-arbiter element that gives very low latencies [49]. Similar techniques can be used with a tree-arbiter element built around a standard mutex, although the resulting circuit, illustrated in Figure 5.9, is slightly slower (four inversions per stage) and is not speed independent.

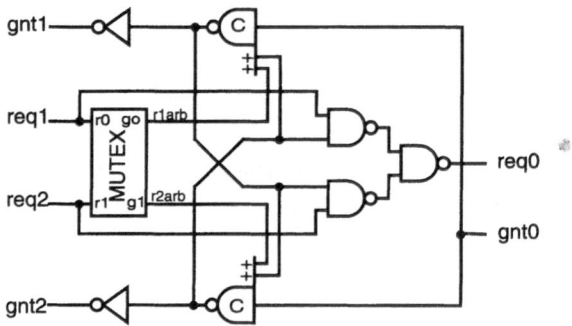

Figure 5.9: Tree-arbiter element, after [49]

This circuit passes input requests through to the output on req0, and determines which event (req1+ or req2+) came first so that, when gnt0 rises, a grant can be given on either gnt1 or gnt2.

Different tree topologies allow the overall *bandwidth* in a busy system to be apportioned as a consequence of the fairness of the mutex element. For example, the tree fragment shown in Figure 5.10a issues equal numbers of grants to all ports, whereas the tree in Figure 5.10b issues 50% of the grants to gnt4, 25% to gnt3 and 12.5% each to gnt2 and gnt1, assuming that all contending devices rearbitrate immediately.

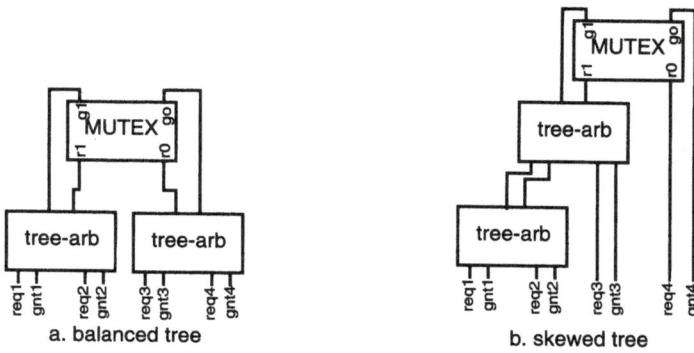

Figure 5.10: 4-way tree arbiters

Sample-prioritise arbiter

Synchronous bus arbiters implicitly use a two stage process of sampling on a clock edge and then prioritizing the active requests to determine which grant to allocate. A similar scheme can be used for an asynchronous arbiter by sampling when an input request is active. This approach scales approximately linearly in hardware requirement with increasing numbers of inputs, but can potentially offer lower latency (which is more important than throughput for the arbiter) than a tree arbiter for larger systems.

Arbitration summary

Arbitration is a key feature of a multipoint channel which can be implemented in a variety of manners, all based around the mutex component. Table 5.1 shows the number of logic inversions involved in an arbitration for the fair-arbitration schemes presented above for an arbiter with "n" contending inputs. (The sample-prioritise approach is not included as it requires a prioritising scheme and is difficult to implement such that it behaves fairly.)

Table 5.1: Arbitration latency

Arbiter	Standard Stage Latency	Number of Standard Stages	Total Latency (inversions)
Single Mutex (n=2)	2	1	2
Mutex Cascade	2	n-1	2(n-1)
Ring Arbiter	22	1 to n	22(1 to n)
Balanced Tree Arbiter	4	$\log_2 n - 1$	$4(\log_2 n - 1) + 2$
Skewed Tree Arbiter	4	0 to n-1	4(0 to (n-1))+2

A balanced tree arbiter was chosen for use in the MARBLE bus described in Chapter 8 because of its low, uniform latency and low hardware costs.

5.6.2 Request drive and hand-over

Each initiator must be able to drive the request of the channel when performing a transfer. Since the arbitration guarantees that only one initiator will use the channel at any one time, the channel-request that is distributed to the address decoder and to the targets is formed through an OR (merge) function of the individual initiator requests. The wired-OR techniques used in backplane buses are not suitable for low-power CMOS implementation, so other solutions must be used.

Centralised request merge
A centralised solution for the request merge function is to use an OR-gate to combine the individual initiator requests to form the channel-request. This would be a more complex XOR-gate if using a 2-phase protocol.

Distributed request merge
The gated approach described above provides the necessary functionality, but is not a modular solution. If true modularity is required, then a scheme using tristate signalling lines may be used. This gives a simpler channel routing requirement (the request and acknowledge wires run with the data wires, with no gates in the path). The problem with such a scheme is that to minimise the risk of malfunction due to the effects of noise and power wastage, the signalling lines cannot be left floating when they are not driven by one of the client devices. Weak charge retention, whilst reducing the power wastage, does little for the noise immunity of the system. To fix this problem, active clamping is required on the lines when they are not in use, along with a controlled drive overlap during hand-overs where both the new and the old drivers together drive the signal to the same value. The hand-over sequence from driver A to driver B is thus:

1. A drives the signal to a known level
2. B starts to drive the signal to the same level
3. There is a short overlap when both drivers drive to the same level
4. A stops driving the signal
5. B drives the signal to its required level

Such a scheme is somewhat more complex than the centralised OR-gate approach described previously.

5.6.3 Push data drive and hand-over

With some variants of the 4-phase protocol there is a risk of drive clashes during hand-over between initiators on a bus channel. This occurs because one initiator could start driving the data lines before the previous initiator has stopped. This would happen if the new driver switch-on and the old driver switch-off had to occur during the same period of the handshake.

Using the early-push protocol (where data is only valid from request+ to acknowledge+) means that the new driver starts driving the push data lines whilst both request and acknowledge are low, holds them driven whilst request is high and acknowledge is low, stops driving them when request and acknowledge are high, and leaves the final phase, when request is low and acknowledge high, for a guaranteed period where none of the initiators drive the bundled data lines. This ensures a clean hand-over, and also fits nicely with the corresponding early-pull data-drive hand-over described in Section 5.7.4, allowing the direction of (some) of the data lines to be changed in subsequent transfers if required.

This approach to data-line drive/validity is equally applicable when using either a tristate or a gate-multiplexed data-path. The former, illustrated in Figure 5.11a, typically allows a smaller implementation, whereas the latter illustrated in Figure 5.11b may be more suitable for synthesis. Both the approaches shown have an "enable" signal associated with each driver. The control logic of the bus interface should monitor this enable signal to determine when the data is driven since its load will increase with wider datapaths. It is probably necessary to include a delay in the monitoring path to provide additional margin and to allow for crosstalk effects as described in the previous chapter.

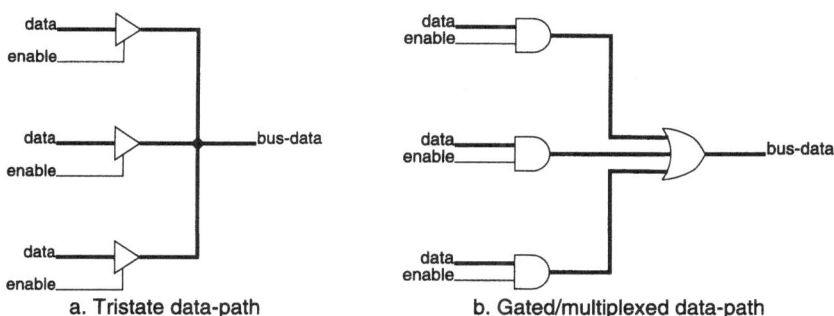

a. Tristate data-path   b. Gated/multiplexed data-path

Figure 5.11: Data-path drive

5.6.4 Transfer deferral/hardware retry

The inclusion of a defer mechanism, where the target can ask the initiator to abandon the transfer and retry it later, allows bridging between multi-point bus channels. To illustrate the necessity of this feature, consider the system fragment shown in Figure 5.12 where two bus channels are connected via a bus bridge. This system allows independent activity on both bus A and bus B when a transfer across the bridge is not required. The bridge allows initiators on bus A to access targets on bus B and vice versa. If both of these actions are required at once, then there is a risk of deadlock: if initiator A0 occupies bus A to access target B3 via the bridge, and at the same time initiator B0 occupies bus B to access target A3 via the bridge, then neither transfer can complete unless the bridge can force either bus A or bus B to become available. The defer technique allows the bridge to achieve this, with the *nacking* arbiter [26,74,81,84] used in the bridge to determine which transfer is allowed to proceed.

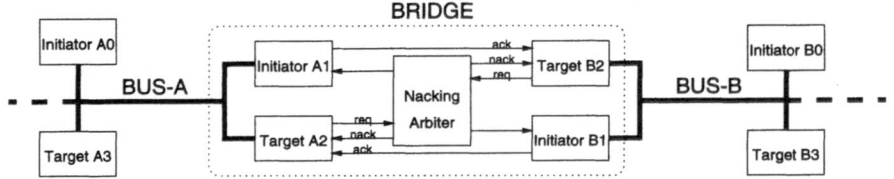

Figure 5.12: Bridging between two bus channels

When a transfer is deferred, the bus-channel cycle must still be completed and the channel returned to its idle state so that it can be used for other cycles and possibly a retry of the same cycle. This action of finishing the cycle, but in such a manner that the transfer is not accepted, is termed a *negative acknowledge* or *NACK*.

Implementation of the NACK requires one additional wire between the initiator and the target, either:

- used as an alternative signalling wire to signal a NACK instead of the ACK, giving a dual-rail encoding of the acknowledge event where one code signifies success and the other failure;

- or used as a bundled data line whose validity is indicated by the ACK signal, in which case the bundled signal indicates whether the transfer was accepted.

The latter approach is used in MARBLE to minimise the wiring overhead of supporting negative acknowledgements. Since these are expected to occur infrequently, the bundled line is (predictively) driven to the ACK state before the actual required value is known, thus optimising for the common case. When a NACK is to be signalled, the prediction must first be corrected causing a slight additional delay.

5.6.5 Atomic transfers and locking

Mutexes and arbiters in general act as hardware interlocks, ensuring mutually exclusive access for a number of clients to a shared resource. The software equivalent, a semaphore, is typically implemented using a read-modify-write activity on a memory location. Special instructions such as the swap (SWP) instruction in the ARM architecture [48] are used by the processor to indicate externally that the memory access should be atomic or uninterrupted by other devices. Any multi-master bus between the processor and the memory must ensure that during such operations other initiators (e.g. other processors) are not granted access to the resource between the read and write cycles.

This can be implemented either by:

- locking the target (as with PCI [72]), so that any other initiator accessing the target is told that it must wait and retry later (see defer above), or;

- locking the bus (as with PI-Bus [64] and AMBA), so that no other devices can access the bus, hence cannot access the target.

To prevent system deadlock, the defer action must be able to override the arbitration lockout; bridges and other components should ensure that only the first transfer of an atomic sequence is deferred. An STG fragment representing this priority of defer over lock is shown in Figure 5.13.

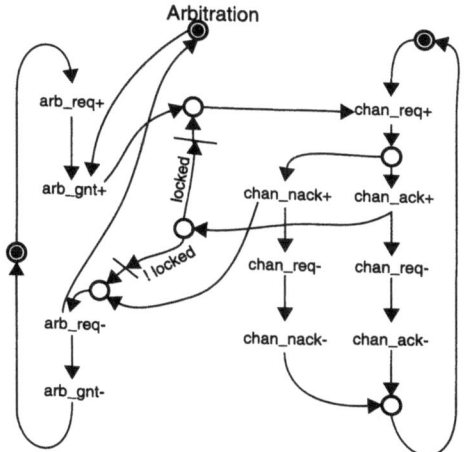

Figure 5.13: Interaction of lock and defer with hidden arbitration

Each of the above approaches has disadvantages. The former requires any other initiators trying to access the same target to poll the target, a situation which is probably best avoided in a low power environment, and requires extra complexity in the target interfaces to implement the locking; the latter may harm overall system performance because the bus could be locked for a long time, meaning that all other initiators are starved of bus activity during the whole period of the locked transfer.

The implementation of these approaches for the asynchronous multipoint channel entails:

- The initiator holding the arbitration request asserted throughout the locked cycles if using hidden arbitration. This means that other transfers are prevented, even if they don't involve the target(s) on which atomic activities are to be performed.

- Choosing to lock either the whole target, or only a part of its address space (e.g. one row of a DRAM) and allowing other initiators to access the non-locked regions when using the target-locking scheme. In either case, extra bus wiring is required to carry an identifier that is unique to each initiator, so that the target can check if an initiator is the one which locked the target.

The hardware required for each of these approaches is identified in Figures 5.15 and 5.16 in Section 5.8 which show complete circuits for an initiator and a target interface.

5.7 Multiple targets

When a channel is connected to more than one target, extra functionality is required to determine which target should respond to each cycle and to ensure that drive and signalling clashes between targets do not occur.

5.7.1 Acknowledge drive and hand-over

Where the initiator is responsible for driving the request signal and arbitration is required to avoid possible drive clashes between initiators, the target is responsible for driving the acknowledge signal. The address on the channel is decoded to indicate that one unique target should respond, thus there is no need for arbitration. The case of broadcast write is the exception, in that it allows the same payload to be written to all targets, and thus requires an acknowledge from each and every target. The techniques used for the acknowledge drive and hand-over are basically the same as those for the request signal as described in the following sections.

Centralised acknowledge merge

A centralised solution for the acknowledge-merge function (in the absence of broadcasts) is to use an OR-gate to combine the individual target acknowledges into the bus-channel acknowledge signal. This is a more complex XOR-gate if using a 2-phase protocol. It is interesting to note that for a broadcast operation, the bus-channel acknowledge can be formed using a C-element to gather acknowledge events from all of the targets before passing them onto the central bus channel. A suitable circuit catering for both broadcast and non-broadcast transfers is shown in Figure 5.14. (The dotted connection to the "-" input of the C-element ensures that the AND-gate output falls before the output ack signal is lowered. In practice this connection and C-element input would be omitted, since the required behaviour is assured by a safe timing assumption.)

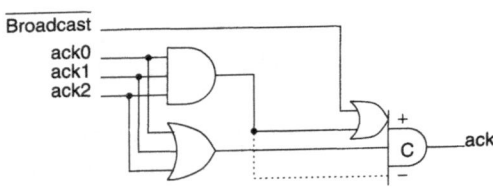

Figure 5.14: Centralised 4-phase acknowledge merge

This circuit passes acknowledges from a device straight through to the central acknowledge during a non-broadcast operation, but when a broadcast is performed all input acknowledges must be high before the central acknowledge is raised, and they must all fall before the output acknowledge falls.

Distributed acknowledge merge

A distributed acknowledge merge function can be created along the same lines as the distributed request merge function in Section 5.6.2. Instead of using the AND function of the arbitration request and grant signals to indicate when to start and stop driving the signal, the AND function of the request and target-select signals should be used. A broadcast operation cannot easily be implemented when using a distributed acknowledge merge.

5.7.2 Target selection

Each cycle on a bus channel may address any of the channel's targets. Part of the forward-going command information is decoded to determine which target is being selected. The decoding may be performed by:

- Acentral decoder, where the channel has only one decoder which waits until a valid address is on the channel and then indicates using a separate target-select line to the addressed target that it should respond.

- A distributed pre-decoder, which requires each initiator to have an address-decoder and to decode the address before it is passed to the channel. The address can be decoded in parallel with arbitration for the channel. The initiator then drives the appropriate target-select line during its push action on the channel.

- A distributed post-decoder, which requires each target to have a partial-decoder, sufficient to allow it to identify all transfers to which it should respond. The address is thus decoded at the far end of the channel, after transport across it.

The MARBLE bus presented in Chapter 8 uses a central decoder since this gives the smallest implementation and gathers the decode logic into one unit, although it does impede bus throughput and increases transfer latency when compared to using a pre-decode and is less modular than the post-decode.

5.7.3 Decode and target exceptions

An error can arise where the address does not map onto any of the targets. This must be handled to avoid a system deadlock. It can be handled in one of two ways:

- holes in the address-map are decoded to address an *error-target* whose job it is to respond indicating that an error has occurred;

- a timeout can be used to reset the channel.

In either case the system processor will have to be notified of the exception, an issue addressed in the next chapter.

5.7.4 Pull data drive and hand-over

When a channel has multiple targets, the situation regarding pulled data drive hand-over is much the same as that for push data drive hand-over on a channel with multiple initiators. Again, using an (early/narrow) protocol where the data is only valid between two consecutive signalling events, (acknowledge+ and request- in this case) provides a clean solution whether using a tristate or gate-multiplexed data path.

Furthermore, if push data is valid between request+ and acknowledge+, or pull data is valid between acknowledge+ and request- (although not both during the same cycle), then the same lines can safely be used for transfers in different directions in different cycles, since there will always be a period where the lines are not driven. So, for example, the same lines could be shared for read and write activity.

If a multipoint channel is to support a broadcast transfer mode where an initiator can push data to multiple targets in the same transfer, then special care must be taken to ensure that the targets do not try to return any information (e.g. status bits) during the pull phase as this would lead to multiple drivers on the pull-data lines.

5.7.5 Defer

The defer operation introduced earlier allows the target to ask the initiator to retry the transfer later. The target must decide whether to accept or defer the transfer and it must signal the result, using either a pulled signal or a negative-acknowledge, depending on the implementation as described earlier in Section 5.6.4. As mentioned above, pulled signals (such as the defer status bit) cause problems with broadcast operations, introducing the possibility of a signal having multiple active drivers at the same time.

A broadcast operation could be supported by mandating that broadcast cycles shall not be deferred (in which case all targets would drive the defer signal to the same level - a safe situation), although this compromises the support for bridging to another multipoint bus. Alternatively a negative-acknowledge approach would allow the central bus control to override ack with nack (as part of its signal merge function) if any of the targets deferred the transfer. There is then the problem of notifying this to all of the targets. The MARBLE bus (and many others) sidestep these problems by not supporting broadcast operations.

5.8 Multipoint bus-channel interfaces

The issues discussed in this chapter affect the design of the centralised bus control logic and the interfaces used to connect devices to the channel. This section brings together these issues to show circuits for two such interfaces.

Figure 5.15 shows the construction of an initiator interface. The initiator device would connect to the left of this circuit, and the multipoint channel to the right. The control logic involved in handling the arbitration pipelining, arbitration locking and defer are highlighted and the connection of the control logic to the tristate datapath drivers is also shown. The "bc_control" lines carry the address, initator identifier, lock signals and the pushNpull signal to the target. The bc_payload lines are then used either to push or pull the data payload in the direction as indicated by the pushNpull signal.

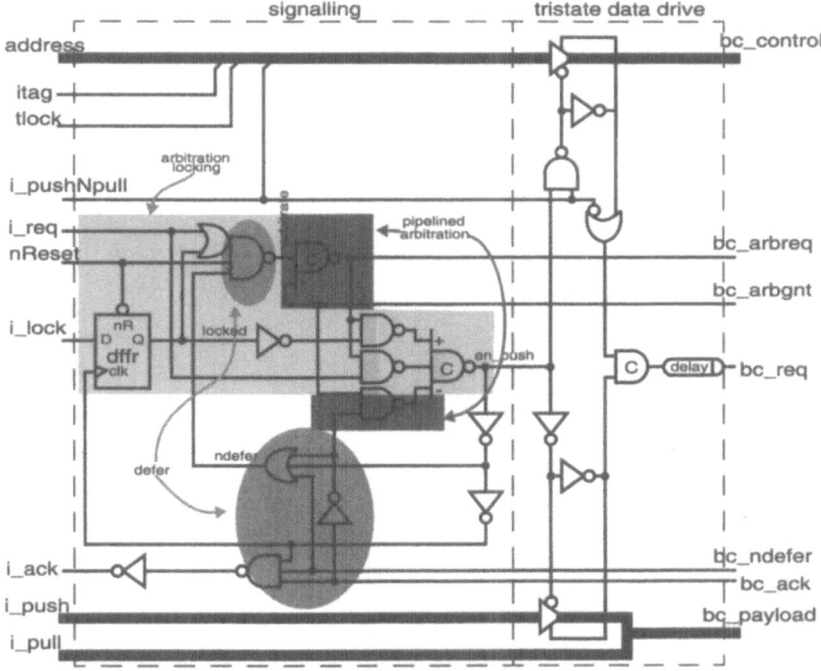

Figure 5.15: Multipoint bus-channel initiator interface

Figure 5.16 shows the corresponding multipoint bus-channel target interface which connects to the multipoint bus channel on the left and the target device on the right, using a dual rail encoded ack/nack approach.

This circuit shows one of the advantages of asynchronous design in that it has been optimised for the common cases:

- When the target is not locked, there is no need to wait until the initiator-id has been checked before deciding whether to accept the transfer.
- The bc_ndefer signal is driven to indicate not deferring before the target device signals its actual requirements using t_ack or t_nack. When not deferring the acknowledge can then be passed straight through with no delay.

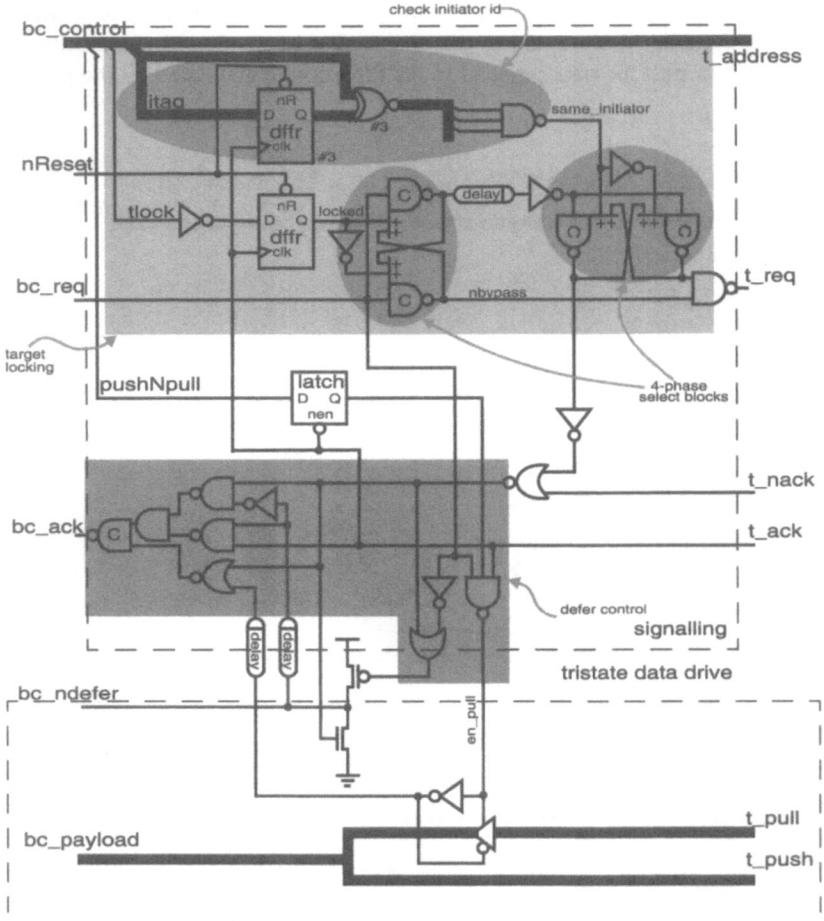

Figure 5.16: Multipoint bus-channel target interface

5.9 MARBLE's link layer channels

This chapter has introduced a range of issues relating to the grouping of many wires to form a multipoint channel, with a defined signalling protocol. To summarise, MARBLE uses a distributed bus approach with the following features for its channel links:

- Single-rail data encoding.
- 4-phase, level sensitive, return-to-zero signalling.
- Centralised request merging and acknowledge merging to ensure reliable operation of the signalling in the presence of noise.

- Tristate data-payload signals which are held driven throughout the relevant push or pull section of the signalling protocol.
- Bidirectional information transfer, using separate wires for the control and payload. The main data payload direction can also be changed from one cycle to the next, allowing the payload to be pushed or pulled.
- Hidden/pipelined arbitration using low latency tree arbiters. All bus grants are therefore early, and the new bus owner cannot start a cycle immediately upon receiving a grant, but must first wait until the bus channel is idle.
- Atomic cycle support through locking of the arbiter.
- Centralised address decoding.

5.10 Summary

This chapter has addressed the design of an asynchronous multipoint channel, concentrating on a 4-phase single-rail protocol. 2-phase signalling is not suitable for use on a multipoint channel because of the problems of drive hand-over. However, if using a short multipoint bus in a kind of centralised hub, then 2-phase signalling may be advisable for long connections between the hub and the client devices.

Whichever signalling approach is used, the issues of arbitration, request-drive, acknowledge-drive and data-drive must all be dealt with, as must the need for a defer primitive if multimaster bus bridging is to be supported. This is further complicated if provision must be made to allow the bus and/or targets to be locked for the atomic transfer support required to implement software semaphores. These are all issues which can successfully be incorporated into a general purpose asynchronous multipoint channel as presented in this chapter.

6. Protocol Layer

The link layer discussed in Chapter 5 provides (one or more) multipoint connections allowing information to be routed between senders and receivers, where these need not be the same from one communication to the next. This chapter introduces the next layer in the hierarchy illustrated in Figure 6.1, the protocol layer. This layer imposes a format onto the use of the channels provided by the link layer. The protocol layer must ensure that the phases of a transfer are performed in the correct order whether these occur on the same channel (for a multiplexed bus), or on different channels (in a demultiplexed arrangement). Particular emphasis in this chapter is placed upon the implications for a dual-channel asynchronous SoC bus, as this can satisfy most requirements.

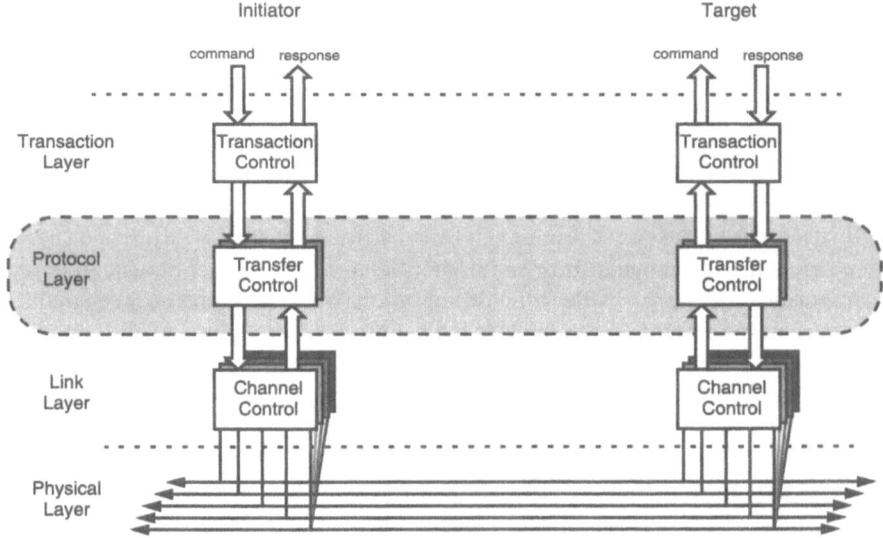

Figure 6.1: Bus interface module hierarchical structure

6.1 Transfer phases

The four distinct phases of any bus transfer are:
- the command phase - indicating what action is to be performed;
- the acknowledge - indicating if the command is accepted;
- the data transfer - transferring the payload between the devices;

- the response - indicating if any errors arose during the transfer.

Each of these phases (of which further details are given below) may occur as a separate communication on the underlying link layer, or they may be combined in some fashion (as discussed in Section 6.4), a decision which affects the performance and hardware resource requirements of the bus.

The protocol layer thus provides a simpler interface to the uppermost layer, the transaction layer, using just two channels per transfer-client, one carrying a command and the other a response.

6.1.1 Command phase

Every transfer begins with the initiator issuing a command to the target. The command indicates:

- the address in the memory map at which the action should be performed;
- the activity to be performed, typically a read or a write;
- the quantity of data to be affected, also known as the transfer size;
- any atomicity requirement, such as locking of the bus or the target.

Other information may be provided as a part of the command, such as the relationship of this command with any previous or subsequent commands. The command may also contain a code indicating which initiator it originated from.

6.1.2 Acknowledge phase

Having received a command from an initiator, a target must either decline or accept the command, and signal its decision during the acknowledge phase. By signalling acceptance of the command, the target is indicating that it can complete the command without causing the system to deadlock. This implies that the target may first need to arbitrate for any shared resources that it needs to perform the transfer, before performing its acknowledge phase. If the target is unable to complete the action because a vital resource (possibly the bus itself if the target is part of a hybrid device that needs to perform a transfer in its initiator role before being able to continue as a target) is unavailable then it should indicate a defer action during the acknowledge phase. The initiator will then retry the whole transfer later.

6.1.3 Data phase

The data phase of a transfer is where the payload is transferred between the communicating devices. For a read transfer, data is transmitted from target to initiator, and for a write transfer data is transmitted in the opposite direction.

Burst transfers may perform a multi-cycle data phase, with a number of data packets being transferred. The command will usually indicate the size of the transfer unit, and may also specify how many such units are to be transferred.

6.1.4 Response phase

The response phase usually occurs at the end of the transfer. It may, in fact, be used by the target to cause the termination of the transfer, especially for bursts that the target can no longer continue. The main use of the response phase, however, is to allow the support of precise exceptions as discussed below.

6.2 Exceptions

The designer of a bus-based environment has to specify what action shall be taken when an error occurs. Typically, this is handled in software by one of the processors in the system, and the bus architecture need only specify how the routine will be invoked. The two common approaches are:

- An *imprecise exception* where an asynchronous interrupt pin of the processor is used to signal the presence of an exception. This carries no information regarding when the problem arose.

- A *precise exception* mechanism passes an extra status bit on each processor interface, so that for every read, write or fetch activity performed by the processor, it is notified of its success or failure. This mechanism identifies uniquely which action caused the error, so that software can better recover from the error.

The imprecise technique is more readily implemented but the precise exception mechanism is more general, and can be used to emulate imprecise exceptions if required. Furthermore, the ARM architecture [48] (and hence the AMULET processors) requires precise exceptions. Further details of the AMULET implementation of precise exceptions are available elsewhere [36,37].

The bus architecture presented here thus uses a precise exception model, always returning a status bit to the initiator indicating the success or failure of the action.

This bit can be affected by bus errors where the address does not map to a known target, and by the target devices themselves where internal errors may be detected. The latter is important since a target may contain multiple logical functions, and may itself be a bridge to a secondary bus.

6.3 Defer and bridging

The potential for deadlock when bridging from one multimaster bus to another was introduced for the multipoint channel in Section 5.6.4. The same problem can arise with a multi-channel bus, when a hybrid device needs to act as an initiator before it is able to behave as a target. Again the solution is to use the defer technique.

The acknowledge phase of a bus transfer is used to allow the target to accept a transfer, or to defer it until later, in which case the initiator will retry the whole transfer beginning with a new command phase. The initiator must thus store sufficient state to be able to restart a transfer until the target signals its acceptance of the transfer, at which point the initiator can discard the stored information.

Where there is a sequencing of the occurrence of the bus phases, for example a fully sequenced bus protocol where the phases occur in strict order (command-acknowledge-data-response), then any phases that would normally occur after the acknowledge phase will be skipped if a deferral is indicated during the commit phase. Similarly, if there is concurrency between the phases, then the content of the data and/ or response phases (if performed) will be ignored and the phase will be retried later.

6.4 Mapping transfer phases onto channel cycles

The flow of activity in a bus transfer was presented above as a sequential stream. However there is no inherent reason why these actions must be performed entirely sequentially. The performance of the bus can be improved through a variety of combinations of the phases to reduce the number of cycles and/or channels required to perform a transfer. Five possible combinations (labelled a to e) are illustrated in Figure 6.2.

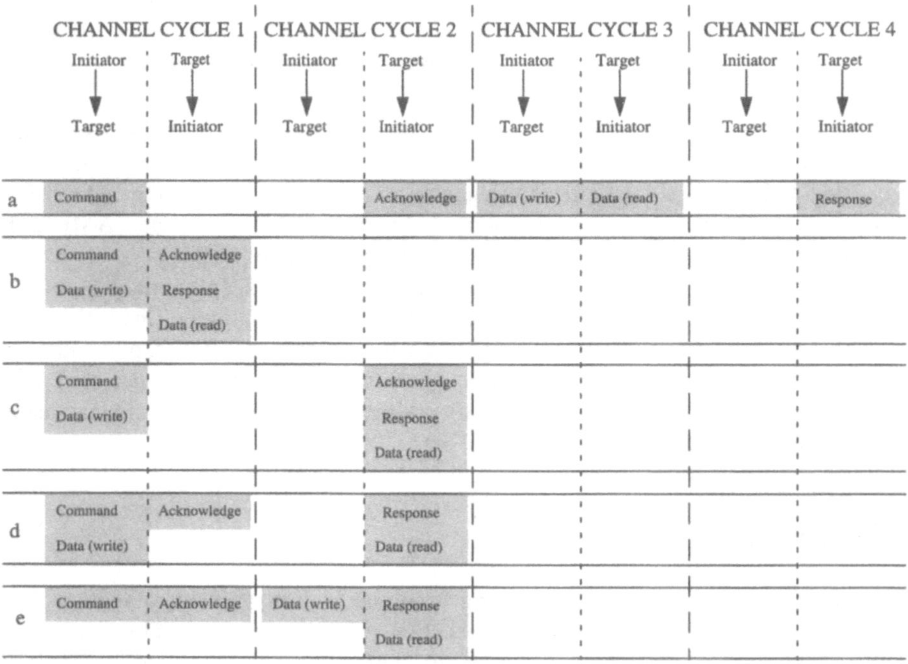

	CHANNEL CYCLE 1		CHANNEL CYCLE 2		CHANNEL CYCLE 3		CHANNEL CYCLE 4	
	Initiator ↓ Target	Target ↓ Initiator	Initiator ↓ Target	Target ↓ Initiator	Initiator ↓ Target	Target ↓ Initiator	Initiator ↓ Target	Target ↓ Initiator
a	Command			Acknowledge	Data (write)	Data (read)		Response
b	Command Data (write)	Acknowledge Response Data (read)						
c	Command Data (write)			Acknowledge Response Data (read)				
d	Command Data (write)	Acknowledge		Response Data (read)				
e	Command	Acknowledge	Data (write)	Response Data (read)				

Figure 6.2: Transfer phase combinations / channel mappings

Mapping a (the first mapping shown in Figure 6.2) shows the sequential approach described above, where each of the phases occurs in a different link layer channel cycle. These may be on the same channel, or separate channels.

Mapping b shows the other extreme case, where all of the phases are performed within the handshaking of a single cycle. This combination is possible because, as shown in Chapter 5, information can be both pushed and pulled in a single 4-phase signalling cycle, although separate wires are required for each of the phases' information.

Neither of the two extremes discussed above leads to both high performance and a low implementation cost (for the reasons described below). A much better solution can be achieved using two cycles, with the transfer phases distributed between them in one of the arrangements shown in mappings c, d and e, described further in the following sections.

6.4.1 Sequential operation using a single channel

The channel cycles required to form a complete transfer could be performed using a channel layer made up of only a single channel, giving a multiplexed bus protocol for the arrangements (a, c, d, e) of Figure 6.2 (above) using more than one cycle per transfer. In such a system the choice of mapping affects the number of wires required since the trade-off being made is time (number of cycles) versus channel-width (determined by the widest combination of phases). Table 6.1 shows how the mapping affects the number of wires, for the five mappings illustrated in Figure 6.2. (In this table C, A, D and R represent the number of wires used to implement the command, acknowledge, data and response phases respectively.)

Table 6.1: Wire requirements if using only one channel for the entire transfer

Mapping	No. of Cycles	No. of Wires	Example C=40, A=2, D=32, R=2
a	4	max(C, A, D, R)	40
b	1	C+A+D+R	76
c	2	max(C+D, A+R+D)	72
d	2	max(C+A+D, R+D)	74
e	2	max(C+A, R+D)	42

Multiplexed buses (using the same wires for different phases of a transfer at different times) are often found at the system board level, where wiring costs are much higher than on-chip and component pins are at a premium. The consequence of the multiplexed approach is the sequential nature of the transfer, which limits the performance. Even for the single-cycle implementation (mapping b in Figure 6.2), the sequential nature of the target device will impose a delay between the presentation of the command and the subsequent return of read data.

For a general purpose macrocell bus, a higher performance is required than can be achieved using sequential phases on a single channel. On chip, the cost of adding additional wiring is reduced (but still significant) and a demultiplexed, multi-channel approach can thus be used.

6.4.2 Parallel operation using multiple channels

A demultiplexed approach gives approximately double the performance of a multiplexed single-channel approach and allows the overlapping of phases from different transfers. Such overlapping is a feature found in most parallel buses that use separate wires for address and data. The advantage of this approach in an asynchronous environment is that the skew between the address and data cycles can be varied whereas in a synchronous environment it is fixed to being a multiple of the bus clock period. This separation onto two or more channels may also reduce the latency added by the bus, whilst incurring little additional cost relative to the single channel approach.

Table 6.2 shows the implementation costs for the mappings shown in Figure 6.2 when each cycle is allocated its own channel. As can be seen, there is little extra wiring cost in using a multiplexed approach, although what is not shown is the increased control logic required for coordination between the channels. In this table C, A, D and R represent the number of data wires used to implement the command, acknowledge, data and response phases respectively, and the H indicates the total number of signalling wires for the channels.

Table 6.2: Wire requirements if using separate channels for each cycle of a transfer

Mapping	No. of Channels	No. of Wires	Example C=40, A=2, D=32, R=2, H=2
a	4	C+A+D+R+4H	84
b	1	C+A+D+R+H	78
c	2	C+A+R+2D+2H	112
d	2	C+A+R+2D+2H	112
e	2	C+A+R+D+2H	80

The four-channel scheme (mapping a above) offers little extra performance over the two-channel alternatives, but requires additional control logic, thus only a dual-channel bus is considered in the remainder of this chapter. MARBLE is such a bus, using the mapping shown in Figure 6.2 e.

6.5 Transfer cycle routing

The command phase of a transfer will either pass an address that can be decoded to indicate which target should respond, or will contain an explicit indication showing the transfer target. In either case the target (or targets for a broadcast operation) is unambiguously identified.

Subsequent phases of the transfer must then be performed between the same initiator and target(s). Much of the complexity of a bus system requiring more than one cycle per transfer (including multiplexed buses) comes from the constraints necessary to ensure such behaviour. Techniques for passing meta-knowledge of which devices should take part in a transfer may be:

- Implicit - where a fixed time relationship (in the form of an interlock, overlap or sequential relationship) between the occurrence of each cycle of a transfer across the bus allows both the initiator and target to switch seamlessly from one phase to the next; or

- Explicit - where each phase is tagged with an identifier (usually unique to the originating initiator) so that all of the phases of a transfer can be routed between the same initiator and target through recognition of the tag by the appropriate unit. The phases may then be performed in a decoupled manner, forming what is known as a split-transfer.

Figure 6.3 shows a timing diagram illustrating typical examples of bus activity for both the interlocked and decoupled schemes. Command A and response D show the activity that would be observed with an interlocked strategy where the command cycle cannot be completed until the response has begun, thus the routing information is implicitly stored in the fact that the two participating devices are active on the command channel when they begin the response cycle.

In contrast, commands B and C and responses E and F show possible relationships that could occur when using a decoupled transfer strategy. Here, an overlap of the end of the command activity with the start of the response activity can still occur but is not a requirement for correct operation. Further, there is no way of knowing just from the signalling shown which command belongs with which response (B with E, C with F, or B with F and C with E), an issue which is investigated further in Chapter 7.

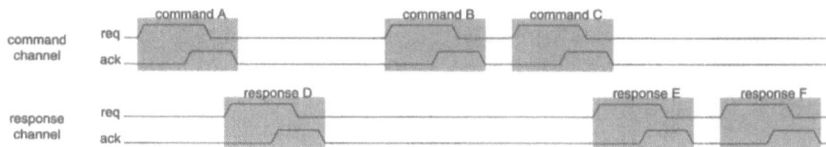

Figure 6.3: Example of dual-channel bus activity

6.5.1 Interlocked protocols

An interlocked bus protocol performs a hand-over of the *routing information* from one phase to the next by ensuring that there is a fixed time relationship between the cycles of the transfer. In synchronous systems, the relationship is often that phases occur in adjacent clock periods. The initiator thus knows to expect a response packet during the clock period after it sent the command.

A suitable equivalent approach that can be used in an asynchronous multi-channel bus is to ensure that there is an overlap of the end of the command cycle with the beginning of the response cycle, either by delaying the completion of the command activity or starting the response cycle before the necessary response information is available for transmission. Ensuring an overlap of activity in this way means that both initiator and target know that activity on the response channel is intended for them.

As stated above, the overlap can be ensured through one of two approaches, either prolonging the command cycle or starting the response cycle earlier than would otherwise be necessary. The example bus activity in Figure 6.4 illustrates this situation, with Figure 6.4a showing the required activity to pass the command and response and Figures 6.4b and 6.4c showing the consequence when using a prolonged command and then an early started response respectively.

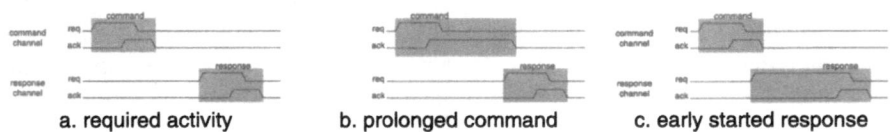

a. required activity   b. prolonged command   c. early started response

Figure 6.4: Causing a command and response to overlap

Using the early started response to provide the overlap is attractive in that it does not hold the command channel any longer than necessary, thus allowing the next transfer to commence (although its response cannot be transferred until after the current transfer has completed). Whilst this strategy works well with initiator started response cycles, it unfortunately degenerates into the prolonged command technique when using a target started response cycle, because this requires the payload to be pushed on the channel yet the response request cannot be raised until the payload is ready.

6.5.2 Decoupled protocols

The alternative to forcing an overlap between the command and response cycles of a transfer is to allow the cycles to proceed in their own time using a decoupled approach.

The advantage of using a decoupled protocol is that it gives a greater overall bus availability, allowing transactions to be interleaved on a cycle-by-cycle basis instead of a transfer-by-transfer basis, effectively bringing the interleaving down from the transaction layer into the protocol layer. Further details on such behaviour, and its uses, are presented in Chapter 7, but what should be noted here is how easily it is achieved when using a decoupled protocol layer.

The disadvantage of using a decoupled protocol layer is the extra implementation cost because the routing information must now be passed explicitly from the command action to the response action requiring:

- an arbitration network to ensure mutually exclusive access to the response channel;

- additional wires (possibly on both the command and response channels) to transmit routing information for the response cycle identifying the initiator from which the transfer originated;

- storage at the target to hold the unique initiator identifier between the command and response cycle;

- an *address decoder* on the response channel to decode the routing information and activate the correct initiator so that it can receive the response.

6.6 Transfer cycle initiation

The final consideration when choosing a protocol layer phase-to-cycle mapping is whether the initiator or target should initiate the response cycle. By definition, the transfer (and hence the command cycle) must be initiated by the transfer initiator but there is no such requirement on the response activity. The implications of the choice of which end should initiate the response cycle vary depending on the type of bus and ordering constraint used.

Figure 6.5 shows the causal relationship between commands and responses, the interlock between the command and response cycles (in bold) and additional arcs (dotted) that reduce the state-space of the system, simplifying the interface controller implementations when using an interlocked protocol.

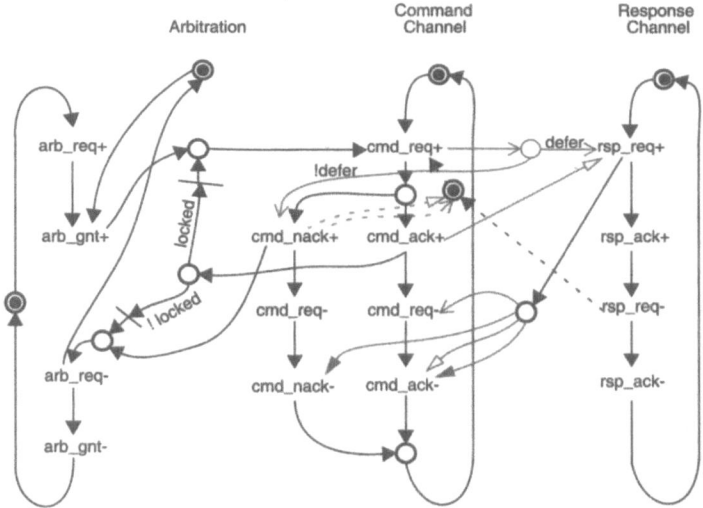

Figure 6.5: Interlocked protocol STG

Three alternative interlocking protocols are shown using the grey arcs in this figure:

- Grey arcs with hollow arrowheads show an initiator started response cycle where the response is not performed for a deferred transfer. This approach avoids wasting response channel bandwidth with redundant response cycles.
- Grey arcs with solid arrowheads show an initiator started response that is always performed even for deferred transfers. This gives simpler control logic since there are no conditional responses.
- Grey arcs with open arrowheads show a target started response that is only performed if the transfer is accepted (i.e. not deferred).

Allowing the transfer initiator to start the cycles (i.e. be the channel initiator) on both channels means that with mapping "e" in Figure 6.2, as used in MARBLE, the write data can be pushed on the response channel as soon as that channel is available, possibly in parallel with the corresponding activity on the address channel. This behaviour minimises the extra write-latency when compared to that of mapping "c" or "d".

When using a decoupled protocol, the choice of which end should start the response cycle is clearer. Here, an initiator started response cycle makes little sense (and is not considered further), since the initiator would then have to either poll the target (leading to unwanted power consumption) or "hog" the response channel. With a target started response cycle on the other hand, the bus is available for use by other transfers during the gap between the target's acceptance of the command and readiness to transmit a response.

6.7 MARBLE's dual-channel bus architecture

MARBLE uses a dual-channel arrangement, where the two channels are known as the command and response channels. The channels are related through a decoupled arrangement as described in this chapter, with all activity on the response channel initiated by the target.

There are thus separate arbiters for each of the channels, and latches at the target to store routing information between the command cycle and the response cycle.

The command channel is used to pass the command phase during the push section of its 4-phase signalling handshake, and to return the acknowledge phase (indicating either deferral or acceptance of the command) during the pull section of the signalling handshake. Deferred commands have no corresponding response channel activity.

Responses are pushed on the response channel when they are available, thus minimising the bus usage for each transfer. All data is transferred on the response channel, with read data pushed at the same time as the exception status bit, or write data pulled later in the signalling handshake. MARBLE thus uses the format described by mapping "e" in Figure 6.2.

The combination of a decoupled protocol with a target started response transfer performed only when the transfer is accepted, restricts the exception support provided by MARBLE. This is because errors detected after the write data has been delivered to the target cannot be signalled precisely (as the exception status is signalled in the same cycle as the write data is transferred, but prior to the write data being pulled to the target) and must instead be signalled using an interrupt. However, such errors are unlikely to occur very often, as they would typically only occur as a result of action involving a bridge from MARBLE to another (possibly off-chip) bus. Two example causes are an address decode failure or a write-data parity failure. Both situations are probably best handled imprecisely since the added delay incurred by the master in waiting for such an exception to be delivered precisely would severely limit overall system performance.

7. Transaction Layer

This chapter addresses the uppermost level of the communication hierarchy of an on-chip bus, the transaction layer. This sits above both the channel layer and the protocol layer, as illustrated in Figure 7.1. As discussed in Chapter 5, the link layer provides (one or more) multipoint channel connections allowing information to be routed between senders and receivers, where these need not be the same device from one communication to the next. The protocol layer, discussed in Chapter 6, is responsible for imposing a format onto the use of the channel layer, mapping transfer phases onto channel cycles such that a bidirectional information transfer can take place between an initiator and a target.

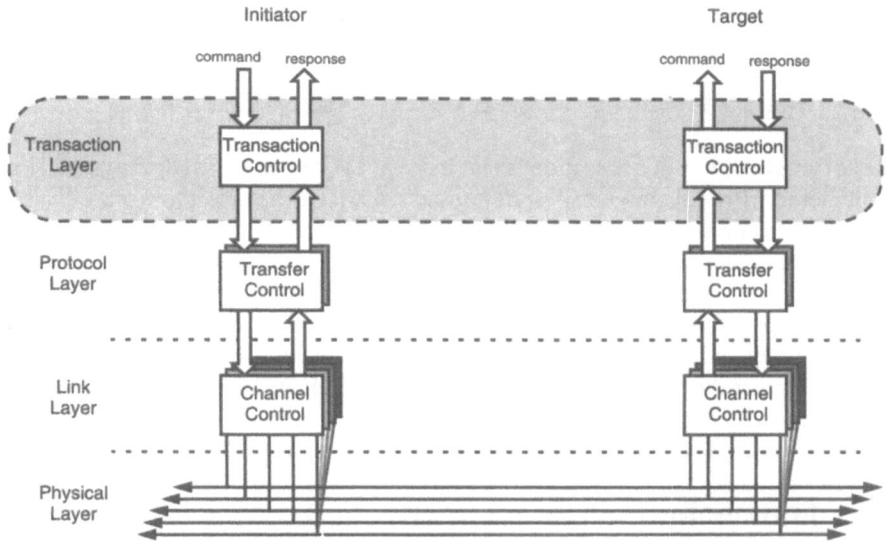

Figure 7.1: Bus interface modules

The role of the transaction layer is threefold:
- to ensure that responses and data are delivered in the correct order;
- to prevent slow targets from unnecessarily stalling the bus, thus giving a greater bus availability for servicing other transactions;
- to allow and regulate the pipelining of transactions across the bus.

These issues are related, and are affected to some degree by the choice of whether an interlocked or decoupled approach is employed at the transfer protocol-layer level.

7.1 Split transactions

A transaction begins when the initiator presents a command (and any write data) to its bus interface, and ends when a response (and any read data) is returned in the opposite direction. Between these actions, the command will have been routed to the target, the target will have performed some action, and then the response will have been returned across the bus.

A system where other bus activity (from other transactions) is allowed between the passage of the command and the response is said to support split transactions, where the command and response are treated as separate packets to be transferred.

Support for split transactions can be provided no matter which implementation technique has been used for the protocol layer, be it interlocked or decoupled, although the issues and implementations involved are slightly different. In fact, with some decoupled protocol layer implementations, every transaction is performed as a split transaction and the problem becomes one of flow control.

The following sections describe how such support can be provided for an asynchronous SoC bus although the same ideas have been used for many years off-chip and have been introduced recently for some synchronous on-chip buses.

7.1.1 Split transactions give better bus availability

The advantage of a split transaction is that it can alleviate the bus occupancy problems encountered with slow targets. A split transaction allows the target to release the bus while processing the command, so that other transactions can use the bus, thus interleaving command and response actions of different transfers on the bus. This is a particularly important issue in the SoC area when the core microprocessor is performing an off-chip memory access. Without split transaction support, the on-chip bus would be occupied throughout the transaction, impeding any other bus activity. With split transaction support, other transfers (including autonomous DMA transfers) can continue unimpeded throughout. The split transfer thus gives greater bus availability and allows the activity of multiple transfers to be interleaved.

7.1.2 Implementation on an interlocked protocol layer

With an interlocked protocol layer a split transaction can be fabricated using one of the techniques shown below. These techniques need not be applied to every unit on a bus, and the split action can be negotiated between the initiator and target on a per transaction basis. In any case, split activity implemented in this manner has a similar hardware cost to implementing a decoupled protocol layer, but may not offer the same availability improvement.

Initiator back-off

The bus can be released before the completion of the target activity related to the transfer through the defer mechanism. The initiator must then retry the command later to check for its completion. In the meantime, the bus is available for use by other initiators. From the outside, this behaviour presents the functionality of a single outstanding command split transaction interface in that multiple transactions can be interleaved on the bus.

If further commands are presented to the target (e.g. from other initiators) while it is busy processing a previously accepted command, then the target must also defer those commands, but may optionally also store them for later processing (though this requires extra hardware and complexity). Similarly, when the target has completed processing a command, it may choose to defer or queue any new commands until it has transmitted the response from the previous transfer.

Generally hardware polling (and hence deferral) is undesirable in a low power system and, although simple to implement, is only used where essential (such as when bridging between multimaster buses). To obtain the increased bus availability through split transactions implemented by a defer mechanism, some buses such as AMBA-AHB add extra connections into the arbitration logic so that the target can indicate when it is ready for the initiator to collect its response. The initiator is then blocked by the arbiter (preventing unnecessary retries) until the target releases it. Clearly this requires additional complexity in the arbitration logic, and this scheme quickly becomes unwieldy (hence the 16 initiator limit on the AMBA-AHB).

Dual role interfaces

An alternative means of achieving split transaction behaviour on an interlocked bus is to provide both initiator and target functionality at each client, as illustrated for the initiator in Figure 7.2.

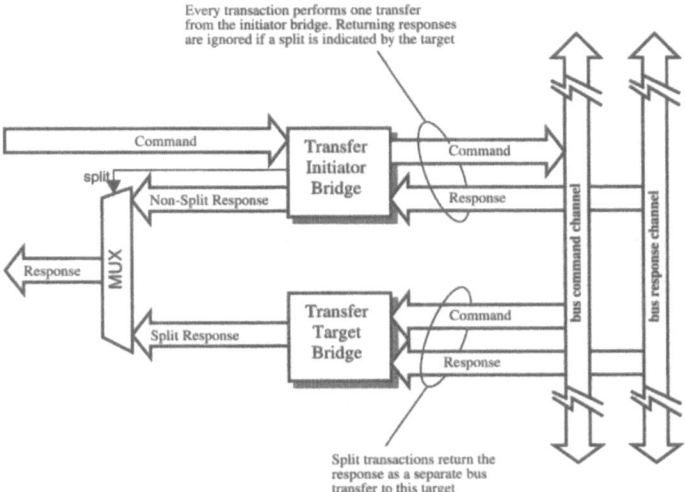

The split transaction can then be performed using two separate transfers; the first with the transaction initiator sending the command to the target, and then the second with the roles reversed so that the transaction target becomes a transfer initiator to send the response (which is received by the transaction initiator acting as a transfer target).

This implementation of split transaction functionality suffers from:

- one extra target loading of the bus for each split transaction capable initiator;
- one extra initiator loading (and corresponding arbitration network support) for each transaction target supporting split operations;
- extra complexity in the address decoding logic since the response transfer must be routed by initiator identifier rather than the address field used to route the command.

As with the polling technique, this approach adds extra complexity, and again it affects the arbiter network, whose performance is critical to the latency experienced by a transfer.

7.1.3 Implementation on a decoupled protocol-layer

A decoupled protocol level, where the target is responsible for starting the response (and read-data) cycle, is inherently a split transaction bus, since every transfer is performed as two discrete actions. All that is required to provide minimal bus occupancy is to ensure that arbitration for the response cycle does not occur until the target is (almost) ready to supply the required information. The bus will then remain available for use.

When the initiator is responsible for starting the response cycles (i.e. pulling the response from the target), the same problem must be addressed as for an interlocked protocol - how to release the bus until it is required again. The solutions are as described previously in Section 7.1.2. This problem is avoided when the target is responsible for starting the response cycles.

For the remainder of this chapter, a *fully decoupled* bus protocol is assumed to mean a bus where the cycles are fully decoupled, with all but the first cycle started by the target device. Any other type of bus is classed as *partially coupled*, including those with an interlocked protocol and those where the initiator is responsible for starting any cycle of a transfer after the first.

7.2 Response ordering

The use of an interlock between the command and response activity of a transfer was discussed in Section 6.5 in the context of passing routing information from one channel cycle to the next. Breaking this interlock to decouple the cycles and give a decoupled protocol was addressed in Section 6.5.2 where additional hardware was introduced to perform this hand-over. The behaviour shown in Figure 7.3, where a second command can be transferred by the bus without having to first wait for a response cycle from the previous transfer, was thus supported, allowing the interleaving of transfers involving different initiators as described earlier.

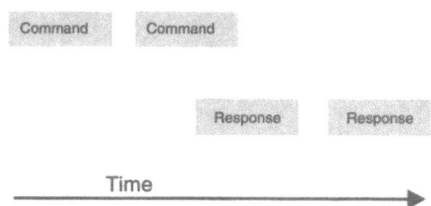

Figure 7.3: Two decoupled commands and two responses

The use of decoupled transfers, either at the protocol level or using the dual role interface technique shown in Section 7.1.2, brings with it an ordering problem. Depending on the speed and behaviour of the (possibly different) targets for each transfer then the ordering of the returned responses will be:

- the same order as the commands were issued, as illustrated in Figure 7.4a;
- different order from that of the commands, as illustrated in Figure 7.4b.

For correct system operation, the latter scenario must be avoided, (as it was with the interlocked protocol layer) or managed through adding a reordering capability at the initiator. The former leads to what is known as a *single outstanding command* constraint whilst the latter approach is said to permit *multiple outstanding commands*.

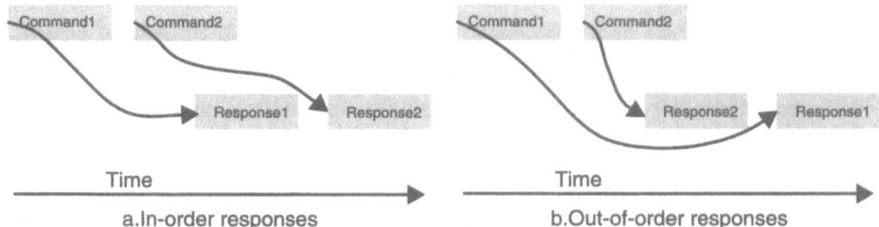

Figure 7.4: Possible order of bus cycles for two transfers

To illustrate the above ordering problem and its solutions it is useful to consider the flow of control during a transfer, as illustrated in Figure 7.5. Consider the token (the black circle) to indicate the primary activity of a transfer (in a bus system with two decoupled channels), which always begins with an initiator sending a command across the bus. Upon reception of the command, there will be a delay while the command is processed. A response is then sent across the bus to the initiator and the control token is then back where it started. The single outstanding command constraint limits (via the throttle) the number of tokens in the loop to just one per initiator whereas the multiple outstanding command support allows multiple tokens, although an upper limit will be imposed for implementation reasons. The decoupling shown in this figure allows for the skewed activity between the command and response channels. For a single outstanding command constraint it amounts to a latch, but a more complex first-in, first-out (FIFO) buffer is required when multiple outstanding commands are allowed.

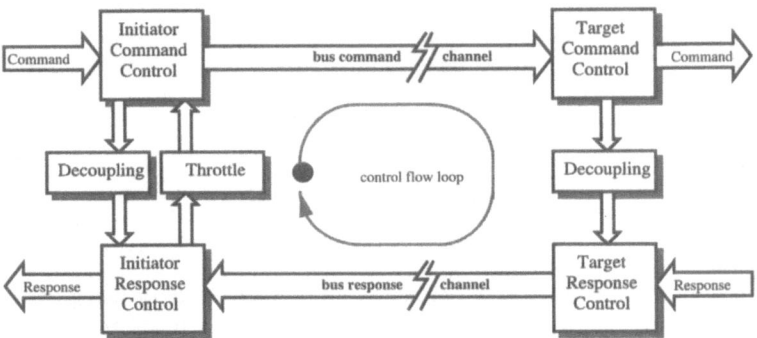

Figure 7.5: Bus transfer control flow

7.2.1 Single outstanding command

The single outstanding command constraint requires that an initiator does not issue a new command until a response cycle has begun for any previous command issued. This prevents responses being presented to an initiator in the wrong order. Note that this approach can still permit the interleaving of transfers from different initiators across the bus, and the responses for these transfers can be passed in a different order from the commands.

Where the interlocked protocol of Section 6.5.1 enforced a constraint (shown by the middle arrow in Figure 7.6) that a command cycle would end only after the start of the corresponding response cycle, the single outstanding command constraint dictates that the next command will begin after the start of the previous transfer's response cycle. This behaviour is shown by the rightmost arrow in Figure 7.6.

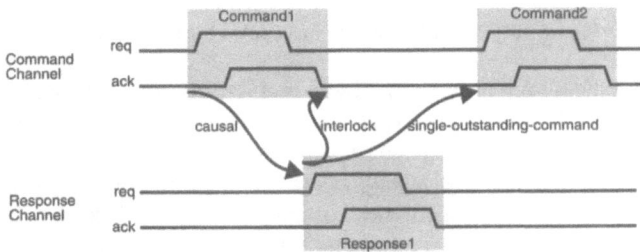

Figure 7.6: Interlocking and single outstanding command constraints

The interlocked constraint also implies the single outstanding command constraint (but not vice versa) hence the need for additional control when using a split transfer protocol.

7.2.2 Multiple outstanding commands and pipelining

Figure 7.7 shows a target interface connected to a target device through a latch in the command path. A latch (as illustrated) allows command n to be processed at the target and to use the response channel whilst the next command, $n+1$ is transferred on the bus. This latch thus provides for a single stage of decoupling between commands and responses, analogous to the pipelining found in many microprocessors.

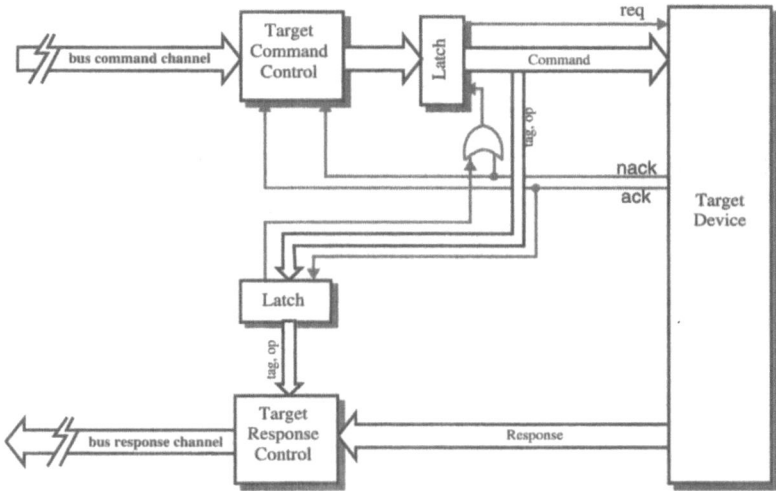

Figure 7.7: Adding pipelining support to a target interface

The signals and the OR-gate drawn in faint in Figure 7.7 show the signalling of the command channel connection to the target device in order to support the defer negotiation. An accepted transfer (indicated using the ack line) places control information (showing the transfer direction and which initiator the transfer originated from) into the lower latch. This information is required when the response cycle is performed. A deferred transfer does not do this.

If the target device contains its own pipeline latches (as would a pipelined RAM, for example), or the latch shown in Figure 7.7 is extended to a multistage FIFO buffer (often termed a command queue in this context) then the target will be able to accept multiple commands before requiring the transfer of responses.

For a single initiator to exploit this extra pipelining it must be able to transmit multiple commands before requiring a response. This means that a split transaction system must overcome, or avoid, the single outstanding command constraint inherent in many interlocked systems, and introduced above for the decoupled transfer protocol.

7.2.3 Number of outstanding commands

One key observation is that, with only one outstanding command, it is impossible for a single initiator to saturate an asynchronous bus. In a synchronous interlocked bus system saturation is possible because the initiator can transmit its next command in the SAME clock period as the response is transmitted, whereas for the asynchronous equivalent there is the small overlap discussed earlier. However, the synchronous system has to include the equivalent time to analyse any wait-state signal within the clock period to determine whether to start the new command, or continue with the old one for another entire clock period.

With higher numbers of outstanding commands, bus occupancy increases until saturation is achieved. The number of outstanding commands required to achieve this varies with the delays of the bus and the target devices. In any case, the number of outstanding commands that can be used is limited by the depth of the target's pipeline.

A final point to note regarding high numbers of allowable outstanding commands is that in order not to stall the bus (and thus significantly impact upon the latency of other bus transfers), targets should be capable of queuing all of the commands sent to them by an initiator. With a single outstanding command constraint, this is not a problem since the bus is unused between the command and response transfers.

7.2.4 A grouping of single outstanding command interfaces

Multiple outstanding command functionality can be emulated by grouping a set of single outstanding command initiator transfer interfaces to form one multi-transaction interface with the usual one command channel and one response channel connections to the initiator device. Extra support logic is required here to distribute commands to the transfer bridges, and gather responses (in the correct order) from these bridges. A suitable arrangement for a four transaction initiator constructed from four single-transfer initiators is shown in Figure 7.8. The AMBA-AHB uses this approach.

This implementation of multiple outstanding command split transaction functionality (in addition to the problems of hardware polling where applicable) suffers relative to a single initiator (and the technique presented later) in that:

- each transfer-initiator must be able to arbitrate for the bus, thus extra latency is incurred in the arbitration stage;
- the extra transfer bridges increase the loading of the bus lines.

The sequential nature of typical initiator devices such as microprocessors and DMA controllers, and the inherent dependencies between the bus transfers that they initiate, mean that the distribution of commands to and collection of responses from the individual single outstanding command initiators within the multiple outstanding bridge must ensure that:

- commands addressing the same target are presented to the target device in the same order as they were presented to the initiator bridge by the initiator device;
- responses are returned to the initiator in exactly the same order as the corresponding commands were presented.

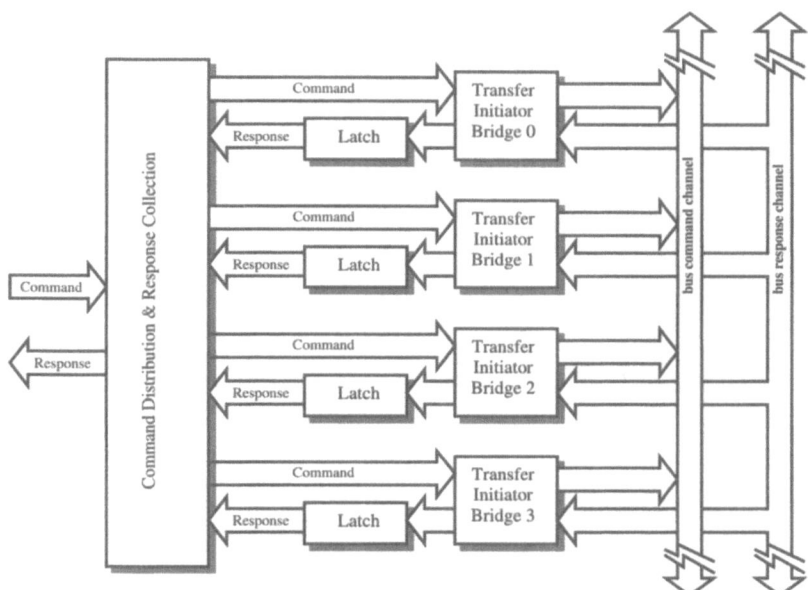

Figure 7.8: Four outstanding command transaction bridge

The latches shown in Figure 7.8 and the response collection logic together form a reorder buffer that may be written to in any order, and read from in a known, fixed order, thus meeting the second requirement.

One technique for meeting the first requirement is for the command distribution unit to issue a command only to a single-outstanding command unit when transfer of the previous command across the bus has begun, although this then limits the interface's ability to saturate the bus. Furthermore, atomic transactions require that the bus be held between transfers, and so the atomic sequence of transfers must either use the same single outstanding command unit or the ownership of the bus must be passed from one to another without releasing the bus. Both approaches add further complexity.

A target capable of supporting more than one transaction at once (thus allowing the pipelining within the target itself) can be constructed in a similar manner, from a grouping of single outstanding command target interfaces. Of course this is only necessary if split transactions are being used to allow pipelining within the target itself, since, even with a simple target, one command can be active in the target, with a second command in transit across the bus.

7.2.5 Sequence tagging and reordering of responses

With a decoupled protocol layer bus (or the provision of similar functionality on an interlocked bus through dual role interfaces as described in Section 7.2) multiple outstanding commands can be supported without the need for interface grouping.

Instead, a single interface is used to pass commands and responses between the bus
and the client, with additional control and data manipulation between the bus interface
and the client.

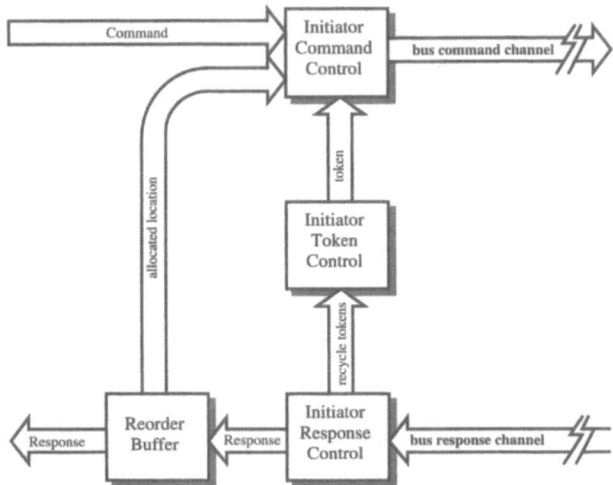

Figure 7.9: Adding a reorder buffer

At the initiator a reorder buffer must be included as illustrated in Figure 7.9 to
allow the reordering of response packets if they arrive out of order (as described
earlier), whilst at the target there may be a need to queue commands between the
interface and the peripheral to prevent the commands waiting for processing from
clogging the bus unnecessarily.

Both the reorder buffer required at the initiator and the low-latency FIFO buffers
used to store commands at the target can be realised using a combination of a counter,
a register-bank and conventional micropipeline FIFO buffers. A low latency parallel-
access FIFO buffer can be built as shown in Figure 7.10a where the two counters must
not *lap* each other. A reorder buffer can be implemented in a similar manner as shown
in Figure 7.10b.

When grouping many single outstanding command initiators to form one multiple
command initiator, each of the subunits had its own connection to the bus arbiter. The
advantage of this approach is that only one connection per (multi-command) initiator
is required to the arbiter.

Where the unique identifier of each transfer initiator clearly showed where
responses should be returned to for the previous approach, their destination implying
the ordering, this approach requires each command to have a unique (from any other
currently outstanding commands issued by the same initiator) sequence tag to hold the
ordering information, and all commands and responses from this initiator will share
the same initiator identifier.

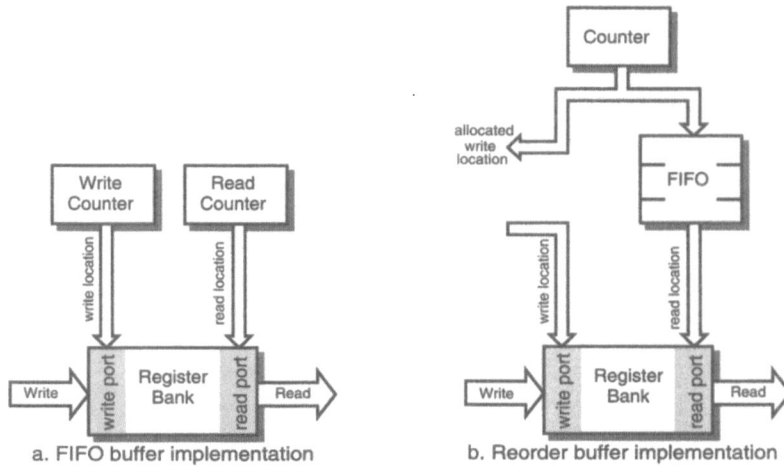

Figure 7.10: Low latency FIFO buffer and reorder buffer implementation

The addition of a command queue at a target that supports deferral (such as a bridge to another bus) requires additional flow control to avoid deadlock. Either:

- The target must agree to accept each transaction before it can be entered into the queue. This ensures that the target has the opportunity to prevent an atomic sequence of transfers from filling the queue and holding the bus if it requires access to the bus in an initiator role before it can process the atomic sequence; or

- The queue must have sufficient free space to hold the entire set of transfers that form an atomic sequence before it can accept the first transfer of that sequence. The queue thus handles defer negotiation on behalf of the target.

If the queue size is much larger than the maximum allowable number of atomic transfers then the second approach is preferable because, with this approach, transfers are only deferred (and hence retried) if the queue is nearly full, a situation which should, hopefully, occur infrequently. A small queue, or the first approach above, gives less decoupling between the activity of the target and the bus, so the likelihood of deferrals if the target requires use of the bus in an initiator role is greater.

7.3 MARBLE's transaction layer

MARBLE uses a decoupled protocol layer, thus providing native support for split transactions without requiring the use of any of the techniques shown in Section 7.1.2. To keep the complexity of the interfaces down for the first AMULET3H chip, the implementation of MARBLE presented in this book permits only one outstanding command per initiator, thus avoiding any reordering or queuing requirements. This does however mean that a single initiator cannot saturate the bus. Further details of the design of the MARBLE bus interfaces are contained in the next chapter, and a complete set of schematics can be found in Appendix A.

8. MARBLE: A Dual-Channel Split Transfer Bus

Chapters 4 to 7 have introduced the complex issues involved in the design of a multimaster SoC bus presenting asynchronous solutions at each stage. The Manchester Asynchronous Bus for Low Energy (MARBLE) described in this chapter provides a concrete example of the feasibility of asynchronous SoC interconnect. It was originally developed for the AMULET3H telecommunications controller chip described later in the chapter, but is intended to be a general purpose SoC shared bus architecture exploiting the benefits of asynchronous design. The chapter is thus broken into three parts:

- MARBLE signal and protocol summary bringing together a complete set of choices from those presented in the earlier chapters;
- descriptions of single outstanding command initiator and target bus interface implementations as used in the AMULET3H telecommunications controller chip;
- an overview of the AMULET3H telecommunications chip illustrating the role of MARBLE within the system.

8.1 MARBLE protocol and signal summary

The fundamental issues relating to the MARBLE architecture have been presented in the previous chapters of this book and are summarised here to show the complete bus architecture.

8.1.1 Two channels

A MARBLE bus consists of two asynchronous multipoint channels as presented in Chapter 5. One of these channels carries the command from the initiator to the target, returning either an accept or defer status. The other multipoint channel carries a response from the target to the initiator (and the read or write data in the appropriate direction). Signal definitions for these two channels are shown in Tables 8.1 and 8.2. Additional signals for connection of the bus interface units to the central address decoder and arbiters are shown in Table 8.3.

Table 8.1: MARBLE command/address channel signals

Name	Function	Description
MAR	Address Request	The command/address channel request line driven by the initiator
MAA	Address Acknowledge	The command/address channel acknowledge line driven by the target
MAO	Address Operation	The command/address operation indicating if the transfer is a read or a write
MAT[1:0]	Address Tag	The command/address tag indicating which initiator the transfer originated from
MA[31:0]	Address	The 32-bit address, driven by the initiator
MSIZE[1:0]	Transfer Size	The size of the data packet to be transferred, which may be byte, half-word or word
MP[1:0]	Privilege Code	Privilege information that may be used by a bus protection unit.
MS[2:0]	Sequential	Sequentiality indicator
ML	Lock	Lock signal indicating that the current and next transfers from this initiator must be performed as an atomic sequence.
nMDEF	Defer	Defer completion of transfer. Initiator should try again later. Driven by target in pull phase of address cycle.

Table 8.2: MARBLE response/data channel signals

Name	Function	Description
MDR	Data Request	The response/data bundle request line driven by the target
MDA	Data Acknowledge	The response/data bundle acknowledge line driven by the initiator
MDO	Data Operation	The response/data operation indicating if the data cycle is a read or a write
MDT[1:0]	Data Tag	The tag indicating which initiator the response/data cycle is destined for
MD[31:0]	Data	Used to send (push) read data on the response/data channel Also used to send (pull) write data on the response/data channel
ME	Abort Error	Abort indicator driven by the target

Table 8.3: MARBLE support signals

Name	Function	Description
MNRES	Active Low Reset	Global bus reset signal
MAarbreqn	Address Arbitration Request n	Signal from initiator to bus arbiter indicating that initiator n requires access to the address channel of the bus
MAarbgntn	Address Arbitration Grant n	Signal from bus arbiter to initiator n indicating that it has been granted access to the address channel for the next cycle
MASm	Address Select m	Signal from the address decoder to target m indicating that it is to respond as the target of the current address cycle
MDarbreqn	Data Arbitration Request n	Signal from target to the bus arbiter indicating that target n requires access to the response/data channel of the bus
MDarbgntn	Data Arbitration Grant n	Signal from bus arbiter to target n indicating that it has been granted access to the response/data channel for the next cycle
MDSm	Data Select m	Signal from the data decoder to initiator m indicating that it is to respond to the current data cycle

8.1.2 Split transactions

The two channels are used in a decoupled transfer scheme with loose coupling between the channels to implement split transactions. One (or more if deferred) command channel cycle and one response channel cycle are used for each transaction as discussed in the previous chapters. All activity on the response channel is initiated by the transfer target. Arbitration for the response channel normally does not occur until the target device is ready to perform the transfer, so as not to impact bus availability for other transfers, although targets are permitted to arbitrate in advance of read-data readiness (i.e. arbitration can take place in parallel with target device activity) to minimise latency.

The AMULET3H chip supports only one outstanding command per initiator, thus avoiding any queuing or reordering requirements. (Additional signals would be required in each channel to convey a *sequencing colour* if multiple outstanding commands are to be supported).

8.1.3 Exceptions

MARBLE supports precise read exceptions by passing an exception status bit (ME) as a part of the response/data cycle. This bit is pushed using a target started response channel cycle, with the consequence that precise exceptions due to a write-data parity failure cannot be supported. Nor can the bridging of write-cycle exceptions to another multimaster bus be supported. Neither of these are required in the AMULET3H system.

8.1.4 Arbitration

A MARBLE bus requires two separate arbiters, one to control initiator access to the command channel, and one to control target access to the response channel. In both cases, because the arbitration is hidden in busy systems (as described in Section 5.6.1), low latency is the key requirement of the arbitration system. In the AMULET3H system there are four initiators and seven targets as described later in this chapter. A 4-way tree arbiter is thus used for the command channel and a 7-way tree arbiter for the response channel. The tree-arbiter elements used for the bus arbiters are built around a standard MUTEX component and each bus client has its own dedicated request/grant channel for interaction with the central arbiters.

8.1.5 Atomic transactions and locking

MARBLE supports atomic transaction sequences through initiator locking of the bus. Thus defer functionality is only necessary in systems supporting bridge-type behaviour where a hybrid device needs to be able to preempt an incoming target access in order to perform an initiator action.

8.1.6 Burst optimisation

Burst actions, where one command can cause the transfer of multiple data-packets (as found in many synchronous buses), are not supported by MARBLE. Instead there is a strict one-to-one correspondence between commands and responses.

This one-to-one correspondence has minimal impact on the overall bandwidth of the system since MARBLE's two channels each have a similar maximum throughput. Further, the implicit knowledge that is transmitted as part of a burst command in other systems is explicitly transmitted in the command with each cycle on MARBLE. The sequential relationships between cycles are transmitted using the three MS[2:0] wires of the command channel. MS[0] is used to indicate if the current address is sequential to the previous address issued by the initiator. MS[2] and MS[1] are used to represent one of the four predictions shown in Table 8.4. The target must mask the MS[2:0] values to ensure that a non-sequential cycle is indicated for the first cycle after a change in initiator (which can be detected by checking the initiator identifier (MAT) with that used in the previous transfer).

Table 8.4: MARBLE sequentiality codes

MS[2:1]	Relationship	Typical applications
00	None	non-sequential memory access
01	Next address LIKELY to be sequential to the current address	instruction fetch ARM load/store multiple
10	Next address WILL be within the same 2^n size memory page (where n is system dependent)	DMA transfer, cache line fetch (wrap-around)
11	Next address WILL be sequential and within the same page	cache line fetch (sequential)

8.2 Bus transaction interface implementation

Interface modules supporting the full range of features described previously have been constructed for the MARBLE architecture. These are based on the multipoint channel initiator and target modules presented in Chapter 5. Schematics for the modules used in the AMULET3H chip are contained in Appendix A.

The initiator and target device connections of these interfaces present and accept the more conventional 4-phase push protocol channel that designers (notably the AMULET group) are accustomed to working with. In each case, two channels are used as shown in Figure 8.1, one for information flowing onto the bus and one for information flowing off the bus. The channels are named in the same manner as their multipoint counterparts with the "-on" or "-off" indicating their direction relative to the bus. There is an exception to this rule: the defer negotiation uses an ack/nack approach on the target channel coming off the bus. There is one cycle on each of the command and response channels connected to a bridge for every non-deferred transfer involving that bridge.

The bundled information on the command on/off channels consists of:

- an initiator-id-tag, a hard-wired unique value for each initiator indicating from which initiator the transfer originates;
- an opcode indicating the type of transfer to be performed, i.e. read or write, and the size of the transfer (MARBLE is a 32-bit bus with possible transfer sizes of 8, 16 or 32 bits);
- an address in the memory map at which the operation should be performed; this will also be decoded as part of the routing action of the transport layer;
- a lock signal indicating that the current and subsequent transfers should be atomic;
- write data, valid only during a write transaction;

whilst that on the response on/off channels consists of;

- an error-status indicating the success or failure of the transfer;
- read data, valid only during a read transaction.

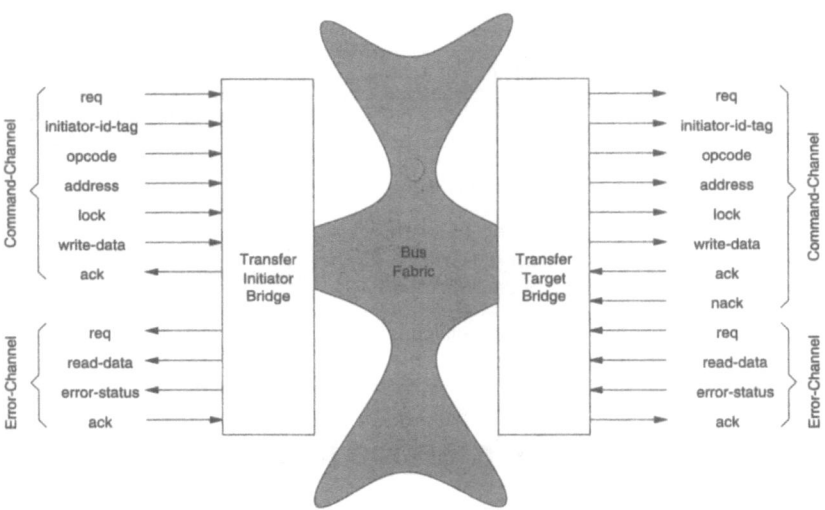

Figure 8.1: Bus bridge interfaces

8.2.1 Interface structure

The design of the MARBLE interfaces presented here can be subdivided into a number of bus protocol control units and data manipulation blocks. The modules presented here all share the same token-flow control structure which was introduced in Section 7.2. The structure of the initiator and target interfaces is shown in Figure 8.2. The dark background in this figure shows the boundaries between the command channel and response channel, and the arrows show the datapaths (wide arrows) and signalling request directions (narrow arrows).

The channel controllers are based on those presented in Chapter 5. The command channel controllers handle the locking and deferral of transfers (there is no response activity if deferred). All locking is performed by the initiator locking the bus, and so the target command controller is very simple. The response channel controllers, in addition to conveying a response, also have to transmit data in the direction indicated by the MAO/MDO bits of the command/response channel. The response channel controllers must, of course, first wait until the appropriate data is available. The following subsections give further details of the implementation of the other blocks shown in Figure 8.2.

8.2.2 Data storage and manipulation

The single outstanding command constraint means that the data-storage requirements of a MARBLE bus interface can be satisfied with simple single-stage latches and the need for FIFO buffers and reorder buffers is avoided.

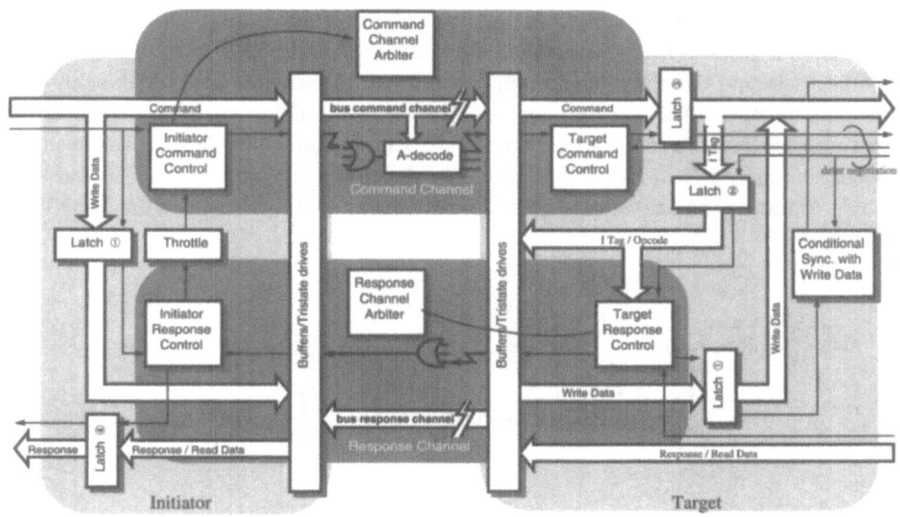

Figure 8.2: MARBLE initiator and target bridge structures

Much has been written on the subject of latch control in asynchronous VLSI design for both 2-phase and 4-phase control. Liu [53] presents a collection of latch controllers with different operating modes and input and output protocols. Bus interfaces provide a good example of where normally-closed latches (which are opened to allow the new data to propagate through, and then closed again before acknowledging the input) can reduce power consumption. They do this by preventing clients from observing activity on the bus data lines unless the client is involved in the transfer, although there is a performance penalty [52] when compared to operating the latch in a normally-open mode.

The majority of latches in the MARBLE interfaces are thus normally-closed, with controllers based upon the long-hold normally-closed latch controller illustrated in Figure 8.3a. This style of latch controller interfaces an early 4-phase signalling protocol channel (the "Ri/Ai" signals) to a 4-phase broad protocol channel (where the bundled signals are valid from Ro+ to Ao-) decoupling the input from the output. Its operation is such that:

- When "Ri" rises, the "Nlt" latch control signal is raised to open the latch.
- Once the latch is open, signal "na" falls causing the output request to be asserted and the "Nlt" signal to fall, thus closing the latch.
- Only once the latch is closed is the input acknowledged. The return to zero of the input channel can then occur in parallel with the cycle on the output channel.
- A new input cycle is held-off from opening the latch again until the previous output cycle has completed, with both "Ro" and "Ao" low.

The latch controller thus includes a degree of load-matching in that if more latches are driven from the "Nlt" signal, its transitions will be slower, delaying "na" falling. The extra buffer (amplification) stages marked **1** and **2** in Figure 8.3a were included to ensure that the latch remains open for sufficient time to allow the new data to propagate through it and they also contribute to the delay necessary to meet the bundling constraint in the output channel. This latch controller variant was used for the latches marked ① in Figure 8.2. The circuits shown in Figure 8.3b, 8.3c and 8.3d are all variants on the normally-closed latch controller that were used in the MARBLE interfaces.

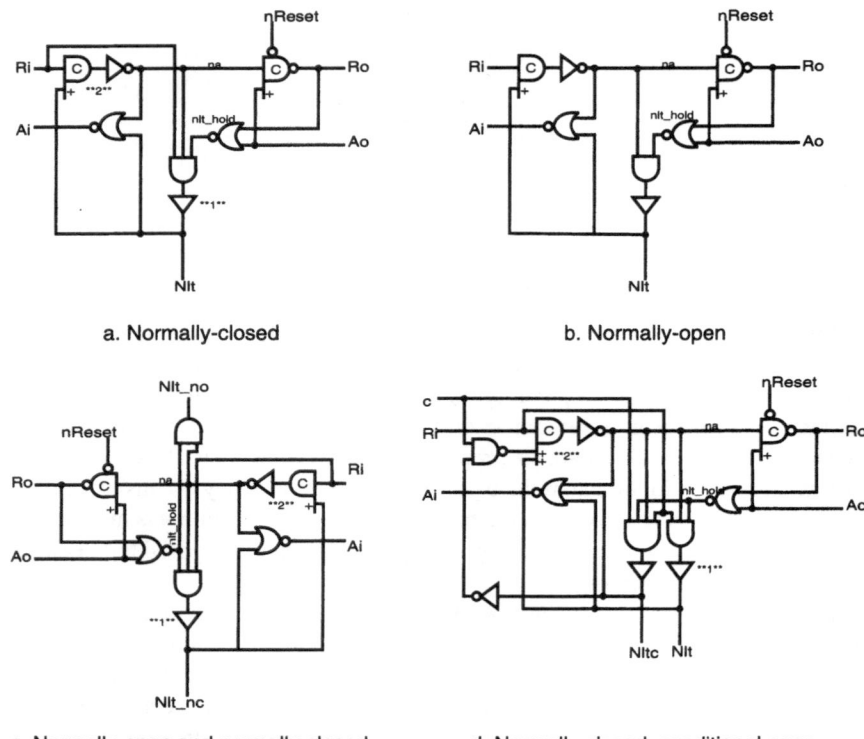

a. Normally-closed

b. Normally-open

c. Normally-open and normally-closed

d. Normally-closed, conditional-open

Figure 8.3: Long-hold latch controllers

Figure 8.3b shows the normally-open latch controller, which allows an incoming request to be propagated immediately through to the output without having first to wait whilst the latch is opened (providing the previous output cycle has finished and the latch is already open) resulting in an improvement in performance. This type of latch controller was used for the latches marked ② in Figure 8.2 where there would be negligible power saving if a normally-closed latch were used.

Figure 8.3c shows a controller with two latch control signals, one (nlt_nc) to control latches in the normally-closed mode and one (Nlt_no) to control latches in the normally-open mode, both sharing the same input and output channel signalling. This mode of latch controller was used in the MARBLE target for the latch marked ③ in Figure 8.2 to allow the Opcode bit indicating whether a transfer is a read or a write to pass through the (normally-open) latch ahead of the other signals in that bundle. This gives sufficient setup time to determine if a rendezvous with write data must be performed by the time the command has passed through the normally closed latch, thus avoiding adding an additional performance penalty to read transfers whilst the check is made.

The final latch controller, illustrated in Figure 8.3d, again has two latch-control signals, both operating in a normally-closed manner. One of these signals, however, does not open for every input cycle, instead its opening is conditional on one of the bundled input signals. This feature is used to open the data-path latches (marked ④ in Figure 8.2) at an initiator only if the transfer is a read, ensuring that the latches remain closed (whilst the response is latched) for a write transfer, thus avoiding any unstable data values propagating from the bus into the initiator and wasting power.

8.2.3 Token management

With a decoupled protocol split transaction bus the size of the reorder buffer (and/or the width of the sequence-tag field of the bus) determines the upper limit on the permitted number of outstanding addresses for an initiator, as described in Chapter 7. The MARBLE bus described here permits only one outstanding command per initiator and so avoids the need for such complexity.

The throttling of the number of outstanding commands for an initiator is performed by restricting the decoupling between the initiator's command and response channel controllers such that an incoming request on the response channel must be received before a subsequent request on the command channel (or the arbitration for the command channel if this is not part of a locked transfer) is performed.

The token management unit is thus a dataless FIFO buffer (or latch when there is only one token). The tokens really correspond to the buffer being initialised to a full state, so that it will issue n output handshakes before receiving any input handshakes, and from then on will output one handshake for each input handshake. Thus for the AMULET3H implementation of MARBLE, which allows only one outstanding command, the token management unit is a latch controller that is initially primed to give an output.

8.3 MARBLE in the AMULET3H system

The AMULET3H [4] chip is the result of the OMI ATOM project [14]. This is a collaborative project funded by the European Union, one aspect of which is to develop a telecommunications chip exploiting the benefits of asynchronous technology. The chip contains a mixture of clocked and asynchronous components, connected through the MARBLE bus and a simple strobed (single master) synchronous peripheral bus. The interconnect provided by MARBLE in this system is shown in Figure 8.4, illustrating where initiator and target interfaces onto MARBLE are required.

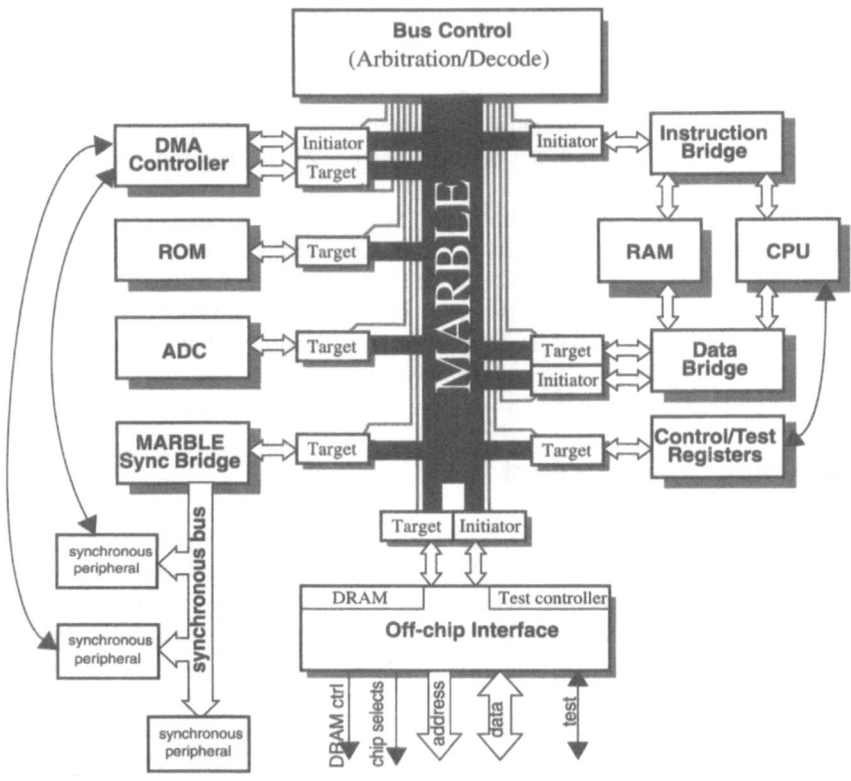

Figure 8.4: The AMULET3H system

The following subsections give short descriptions of the behaviour and function of each of the macrocells in the system, focusing on the interconnect required by each part.

8.3.1 AMULET3 processor core

The core organisation of the AMULET3 microprocessor core is shown in Figure 8.5 (from [4]). The architectural features summarised here have been described in detail elsewhere [4,36,37]. This is the third generation asynchronous microprocessor from the AMULET Group at the University of Manchester and is code compatible with the ARM synchronous microprocessors.

Key AMULET3 core features include:

- a modified Harvard architecture;
- decoupled prefetching with branch prediction;
- support for precise exceptions;

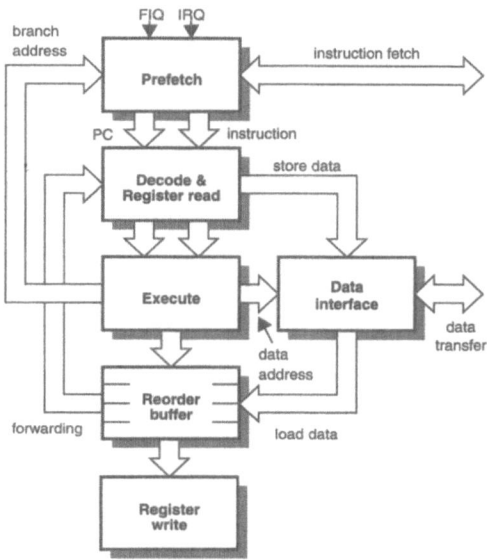

Figure 8.5: AMULET3 core organisation (from [4])

- full support for the ARM architecture v4T [48] including the Thumb 16-bit compressed instruction set;

- register dependency and exception handling using an asynchronous reorder buffer from which results can be forwarded to allow out-of-order completion and speculative execution of instructions;

- a very low power halt mode implemented through suspending the prefetch activity until an interrupt occurs;

- debug support using comparators and mask registers on both the instruction and data ports to provide basic (exact) breakpoint and watchpoint facilities.

Simulations show that the processor core will operate at an average frequency of around 130 MHz on "typical" silicon at 3.3V, 25°C. This is a similar performance to the equivalent synchronous ARM9 microprocessor operating at 120MHz on the same silicon process.

There are two memory interfaces to the processor. One, (the data port), is used for memory accesses by load and store instructions. The other (the instruction port) is used for instruction fetches, and also for data loads to the program counter. This last peculiarity means that the processor does not strictly adhere to the Harvard architecture found in many other high performance processors, but it simplifies the implementation of the processor and allows program counter loads to be performed in parallel with data register loads. The consequence of this design choice is that the memory subsystem must provide a unified view of memory.

8.3.2 RAM

Eight kilobytes of on-chip SRAM are provided to hold speed-critical routines. The RAM comprises eight 1 KB pseudo dual-ported blocks. As illustrated in Figure 8.6 each 1KB block consists of a single ported RAM array with separate line buffers for each of the instruction and data ports, and an arbiter to ensure that only one port uses the RAM array at once. Each block has a RAM array bypass path that allows accesses within the same line as the previous access to the same port to be satisfied from the line buffer.

The combination of the line buffer and the same line access bypasses allow the arbitrated access single ported RAM to achieve almost the same performance as a fully dual-ported RAM (except for when very rare conflicts occur) but with much lower cost.

The RAM blocks are connected via two shared local buses to bridges (see below) that provide connection to both MARBLE and the AMULET3 core.

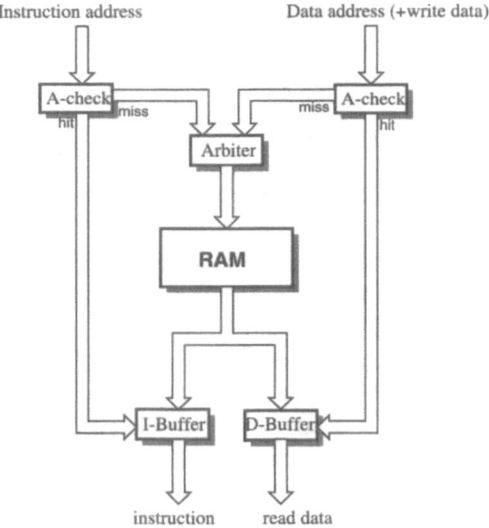

Figure 8.6: 1KB RAM block

8.3.3 ROM

The simplest MARBLE target in the AMULET3H chip is the 16kB on-chip ROM used for program store. This target will generate an exception for any attempted writes to it. The ROM access time (including control logic) is around 4ns making it one of the faster targets on MARBLE; to enhance performance further when running code from ROM, its MARBLE interface issues an early request to the arbiter (i.e. speculatively before the read data has been retrieved) for the MARBLE response channel, so as to minimise the access latency.

8.3.4 DMA controller

The AMULET3H asynchronous subsystem contains an autonomous DMA controller with 32 independently programmable channels, each capable of performing transfers between two memory-mapped addresses that may correspond to memory or peripheral devices. The DMA controller operates using a store and forward technique, requiring two complete MARBLE transactions per DMA transfer. MARBLE does not support fly-past DMA where the controller sets up a direct data transfer between the two peripherals.

Much of the DMA controller was synthesised using Balsa [12], with custom-layout used for the registers and the synchronous DMA request prioritisation logic.

8.3.5 External memory/test interface

A 16-bit data bus is used to interface the chip to external memory parts. In normal operation the AMULET3H chip will be the only master on this bus, all traffic to the external bus passing over MARBLE and onto the bus via a target interface. An on-chip test controller allows the ownership of this bus to be reversed for test purposes, so that it may be used (via a MARBLE initiator interface) to apply test vectors to, and read results from, any peripheral that can be addressed over MARBLE. The external interface can thus be either a MARBLE target or initiator.

The MS field of the MARBLE command is primarily for use by the external memory interface so that it can exploit the behaviour of DRAM parts which typically provide much faster access times for all but the first access of a group of accesses to the same row.

8.3.6 ADC/AEDL

The synchronous section of the chip contains an analogue to digital convertor (ADC). The interface logic that connects this (as a target) to MARBLE permits the reading of the ADC result to be synchronised by waiting for a signal which indicates the result is ready. In the context of this chip, this is known as an asynchronous event driven load (AEDL). The key benefit provided by the AEDL is that it allows synchronisation of the software running on the processor with the synchronous hardware event. The split transaction support of MARBLE means that the processor can be stalled waiting on the ADC without impeding the DMA engine.

8.3.7 SOCB

There are three interfaces between the synchronous and asynchronous regions of the chip, the ADC/AEDL, the DMA controller requests and the bridge between MARBLE and the synchronous on-chip bus (SOCB). The synchronous modules in the AMULET3H chip comprise library modules and custom units from both the

commercial partner in the project and the manufacturer's component library. They
provide the dedicated functions necessary for telecoms applications.

The SOCB is a simple, synchronous, interlocked protocol, strobed peripheral bus
with only one master (the bridge from MARBLE). Figure 8.7 shows a timing diagram
of the operation of the SOCB illustrating both the read and the write behaviour of the
bus. A short description of each of its signals is given in Table 8.5.

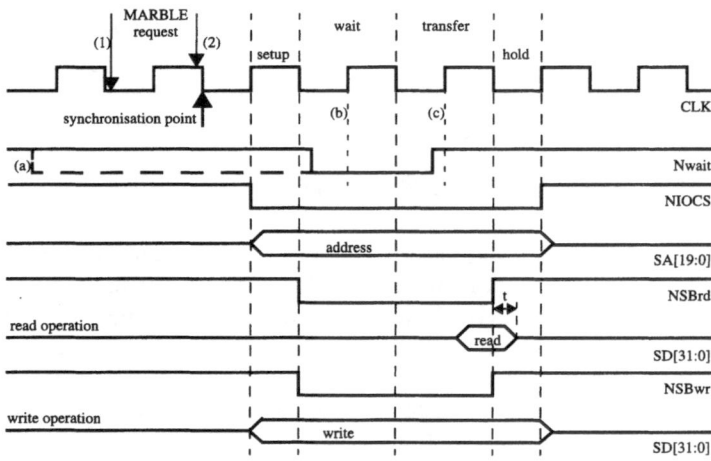

Figure 8.7: SOCB timing diagram

Table 8.5: SOCB signals

Signal	Function
Nwait	wait signal allowing the peripheral to prolong a transfer by another cycle
NIOCS	peripheral access strobe
NSBrd	read strobe
NSBwr	write strobe
SA[19:0]	20-bit address
SD[31:0]	32-bit data lines
SIZE[1:0]	data transfer size (byte, halfword or word)
CLK	bus clock

The SOCB clock runs at up to 66 MHz, with a minimum of two cycles required for
each transfer. There is no pipelining of address and data activity and the address must
be held stable throughout the data transfer.

When an SOCB transfer is requested from MARBLE, a synchronisation of the
request signal must be performed. Since SOCB transfers always commence on the

positive edge of the clock, the synchronisation is performed on the negative clock edge. Hence the delay from the asynchronous request rising to the start of a transfer takes a wor.t case time of 1.5 SOCB clock cycles and a best case of 0.5 clock cycles, giving an average of 1 cycle added latency due to the synchronisation. Figure 8.7 shows the arrival of two asynchronous requests from MARBLE, labelled (1) and (2), leading to the worst case and best case synchronisation times respectively.

On the next rising clock edge after synchronisation, the bridge will drive the address, size and data (for a write transfer) onto the SOCB. An additional signal, NIOCS is also supplied to activate the SOCB address decoder. A half-cycle setup time is then provided before read or write strobes are generated.

If the bridge wishes to read from an SOCB peripheral then the information must be stable at the rising edge of NSBrd. In order to avoid drive clashes, the turn-off time (labelled t) must be less than the hold time.

The Nwait signal may be used by a peripheral to insert a wait-state, allowing the duration of a transfer to be extended. Nwait is sampled on the positive edge of the clock when a read or write strobe is active. If low, the cycle is converted to a wait state. Nwait may be held low for an unlimited number of clock periods.

8.3.8 Instruction bridge and local bus

The instruction bridge connects the AMULET3 processor instruction port to the on-chip RAM and to MARBLE. This unit will never perform write actions to the RAM or to any other MARBLE target, and is optimised for maximum bandwidth. It uses standard unidirectional broad-protocol push channels interconnected as shown in Figure 8.8. The FIFO buffer holds control information to switch the multiplexer to the

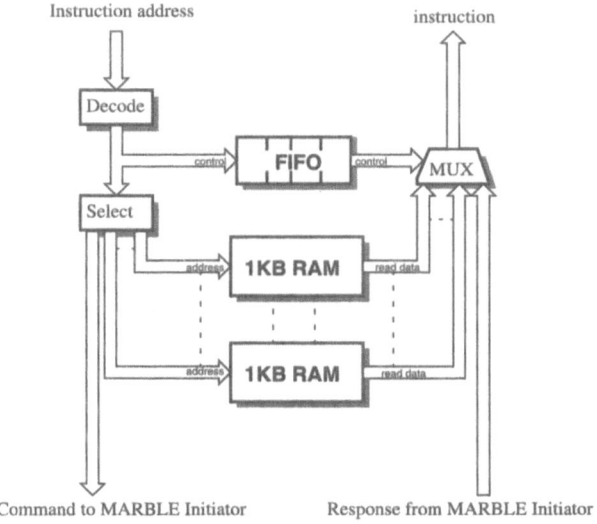

Figure 8.8: Instruction bridge and local bus

correct channel for the returning instruction. The depth of this FIFO buffer imposes an upper limit on the number of RAM blocks that can be in use at any one time. This can be greater than one because of the decoupling between the RAM address inputs and data outputs provided by the RAM line buffers.

8.3.9 Data bridge and local bus

The data bridge connects the AMULET3 processor data port to the RAM and the MARBLE bus as illustrated in Figure 8.9. This unit is more complex than the instruction bridge since it has to support both read and write transactions. It also has to map the ARM swap instruction into a read followed by a write; this is performed in the unit labelled SWP in Figure 8.9. It must also act as both an initiator and a target on MARBLE to allow other initiators, such as the DMA controller and test interface, to access the RAM. In contrast to the instruction bridge, the data bridge is optimised for minimum latency (rather than maximum bandwidth) to minimise processor stalls due to data accesses.

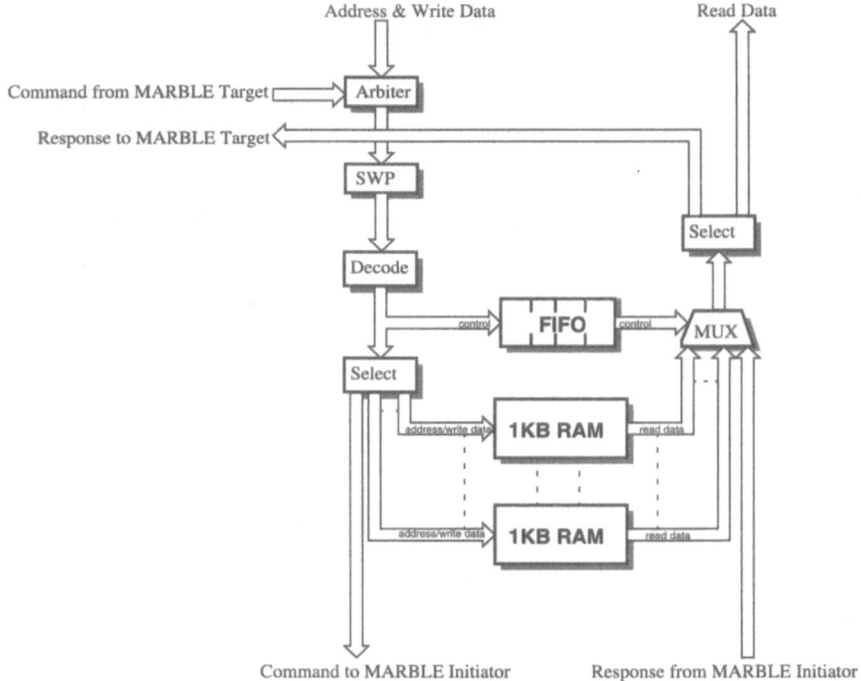

Figure 8.9: Data bridge and local bus

Arbitration is required in the data bridge to resolve the contention that can occur when the processor core and the MARBLE target interface both require service from the RAM. The arbitration is performed before the SWP expansion to simplify the arbiter (which would have to support locking if the expansion was performed first). The structure shown does not support concurrent access to MARBLE for the CPU and to the RAM from the MARBLE target (which would improve throughput) as this would give a longer latency due to the extra control complexity.

A final point to note regarding the connection of the processor core and the RAM to the MARBLE bus via the local bus bridges is that a throttle on the instruction fetch is required in the processor core to avoid deadlock. This is illustrated in the following example: suppose the processor prefetch unit fills the buffer internal to the core, and occupies the local buses and MARBLE whilst fetching instructions from ROM. If the processor tries to perform a data access to the RAM, this can proceed unimpeded, but if a data access to a peripheral on MARBLE is attempted, then the transfer must wait for MARBLE to become available. Normally this would occur as instructions filter through the processor, but if there is a dependency between the next instruction and the data access then the system is deadlocked. The solution is to restrict the number of outstanding instruction fetches such that the returning instructions can be stored at the processor without obstructing the local bus, MARBLE or the RAM.

8.4 Summary

This chapter has introduced MARBLE, a dual-channel asynchronous System-on-Chip (SoC) bus using a decoupled protocol to provide inherent support for split transactions without the need for polling or other complex implementations. This allows the connection of asynchronous macrocells (without introducing a clocked bus) to facilitate the construction of entirely asynchronous VLSI SoCs.

Modular construction details have been provided for single outstanding transaction initiator and target interfaces which have been used in a real system, the AMULET3H. Figure 8.10 shows a die plot of the final AMULET3H layout with an overlay showing the location of the MARBLE bus wires and drivers and other major system components. The control logic for the initiator and target interfaces of the DMA controller can be seen at the leftmost end of the region labelled MARBLE. The control logic of the other MARBLE interfaces is included into the initiators' and targets' compiled blocks of standard cells.

The performance of the AMULET3H MARBLE implementation is evaluated in the following chapter and compared with synchronous alternatives offering similar functionality.

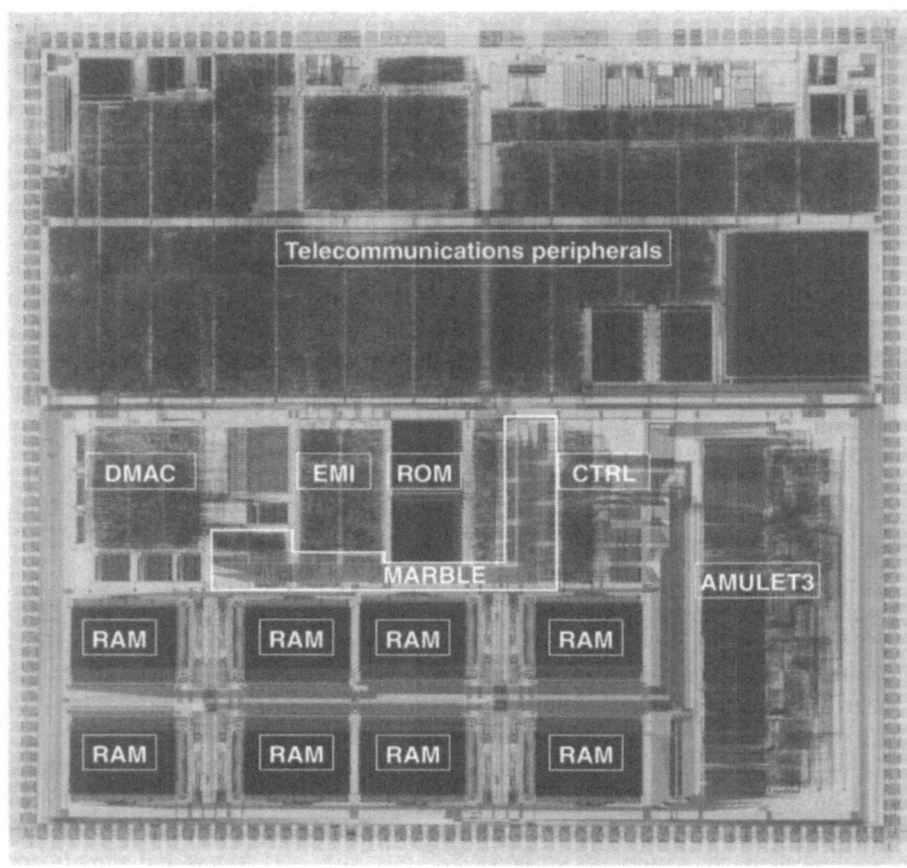

Figure 8.10: AMULET3H die plot

9. Evaluation

Chapters 4 to 7 introduced the issues involved in designing an asynchronous SoC bus. A complete bus was then proposed in Chapter 8 and shown in the context of the AMULET3H subsystem. This chapter presents some of the post-layout simulations performed to validate the functionality of MARBLE and to measure its performance.

The chapter is divided into four sections:

- the MARBLE testbed used to explore the limits of operation of the interfaces and to verify the correct controller behaviour with a very fast environment;
- the AMULET3H simulations used to validate and measure the MARBLE performance in a real system;
- an analysis of delay distribution showing how the different phases of the MARBLE protocol and circuit implementations contribute to the time required to perform a bus transaction;
- performance comparison with common synchronous SoC buses.

All of the simulations used in the comparisons presented in this chapter were performed using EPIC TimeMill v5.2 from Synopsis assuming "typical" silicon parameters at 3.3V, 25°C.

9.1 MARBLE testbed

A standalone testbed with four initiators and eight targets was created to validate the functionality of the MARBLE interfaces prior to their connection within the AMULET3H subsystem and independently from the initiator and target delays that are endured in a real system.

This testbed used initiators that immediately request another transfer as soon as possible after a previous transfer has completed (i.e. they have a very low cycle duration on the non-bus side of the interface), and targets that have very low latencies. These behaviours were achieved by tying back the outgoing requests to their corresponding acknowledges, and feeding incoming requests via an inverter driven from the corresponding acknowledge.

The use of these high performance initiators and targets allowed the testbed to exercise the timing of the interfaces more rigorously than should occur in practice. The testbed was run with two extremes, a short bus (giving lower load and crosstalk effects) and a long bus to allow the interface behaviour to be verified independently from the initiator and target delays that are endured in a real system such as the AMULET3H. In each case the testbed used asymmetric arbiters (as opposed to the symmetric ones in AMULET3H) to obtain minimal arbitration delay.

After verification of the MARBLE interface functionality using the testbed, and minor improvements to the interface circuits, the defer logic was removed and the interfaces were inserted into the AMULET3H system allowing realistic performance measurements to be made as described below.

9.2 Simulation of MARBLE in AMULET3H

Simulation of MARBLE within the AMULET3H subsystem is a matter of loading ARM code into the ROM, on-chip RAM or into a memory model connected to the external memory interface and then simulating the whole system using the TimeMill simulator.

All of the programs were run from ROM as that is the fastest MARBLE target that can be used by the processor core in the AMULET3H subsystem, thus allowing greater bus occupancy than using other, slower targets. In each case AMULET3 booted from off-chip code to initialise the interrupt vectors and configure the external memory interface before running the test program.

The signals visible in the following simulation waveform plots are divided into the following groups:

Group A: MARBLE command channel signals (as described in Table 8.1 on page 84), command channel arbitration request and grant signals (ma_req and ma_gnt); mar_i request signals from the initiators to the MAR centralised OR gate,

mas signals from the address decoder to the targets,

maa_t acknowledge signals from the targets to the MAA centralised OR gate.

Group B: MARBLE response channel signals (as described in Table 8.2 on page 84), response channel arbitration request and grant signals (md_req and md_gnt),

mdr_t request signals from the targets to the MDR centralised OR gate,

mda_i acknowledge signals from the initiators to the MDA centralised OR gate.

Group C: AMULET3 processor core instruction fetch port. Visible signals here are the outgoing address (ia[31:0]) with its request and acknowledge and the returning instruction (id[31:0]) with its request and acknowledge).

Group D: AMULET3 processor core data access port. Visible signals here include the address (da[31:0]) with its request and acknowledge, write data (wd[31:0]) and read and write signals (which all have the same timing as the address). Also shown is the read data (rd[31:0]) and its signalling wires. The exception status bit of this bundle is omitted since it is always inactive in all of the simulations illustrated here.

Group E: ROM access signals between the ROM and its MARBLE target bridge. Visible signals include the address and its request (ardy) and acknowledge, and the read-data bundle (don[31:0], donr and dona).

The MAT[1:0] and MDT[1:0] signals are used to indicate which initiator the address or data cycle belongs to, as explained in Section 6.5.2 for a decoupled transfer protocol. The unique identifiers that can be encoded on these lines were allocated as:

0. External test interface controller
1. AMULET3 processor instruction port
2. AMULET3 processor data port
3. DMA controller transfer engine

The address channel arbitration signals, and the lines connecting the initiator interface to the centralised OR gates for request and acknowledge merging, are shown in the simulation traces as signals ma_req[i], ma_gnt[i], mar_i[i] and mda_[i] respectively where "i" is the unique initiator identifier listed above, as used on the MAT/MDT lines.

Similarly, each target interface has its own connections to the data channel arbiter (signals md_req[t] and md_gnt[t]), from the address decoder (mas[t]) and to the centralised OR gates for request and acknowledge merging (signals maa_t[t] and mdr_t[t]). Here, "t" is allocated as:

0. 8KB on-chip asynchronous SRAM
1. On-chip asynchronous ROM
2. External (off-chip) memory interface
3. Analogue to digital converter (ADC)
4. MARBLE to synchronous bridge (MSB)
5. DMA controller configuration registers
6. AMULET3 processor configuration/test registers

9.2.1 Single initiator to single target

In many scenarios the processor may run code directly from the ROM, so the available bandwidth for a single initiator using the bus (whilst all other initiators are idle) is an important parameter. The maximum throughput for this type of communication was measured by allowing the AMULET3 processor to fetch instructions (a mixture of "move" and "and" instructions) from the ROM which are then subsequently discarded because they fail their condition codes (set by an earlier arithmetic instruction). This behaviour causes the CPU to fetch instructions more rapidly than if the instructions have to propagate through the processor pipeline, possibly encountering dependencies along the way, thus making heavier use of the bus.

The resulting trace is shown in Figure 9.1 where the command and response cycles for five consecutive ROM accesses are highlighted by the overlays. The five command cycles performed between the two cursors take an average (over five transfers) of 18.3ns each, giving a throughput of 55 million transfers/second for the processor instruction port. The figure clearly shows the overlapping of the command cycle of a transfer with the end of the response cycle from the previous transfer. It also shows the idle periods between transfers (when both request and acknowledge are low) as a consequence of the single outstanding command constraint. These actually acount for almost 6ns (a little greater than the access time of the ROM), or around one third of the cycle time.

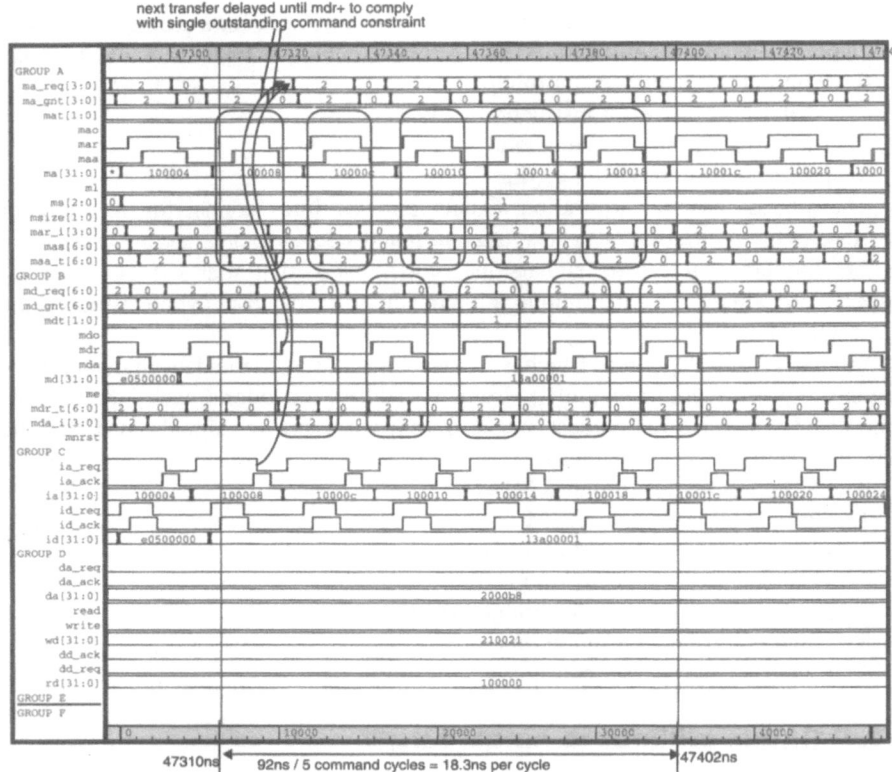

Figure 9.1: AMULET3 fetching code from the ROM

9.2.2 Two initiators accessing different targets

Using only a single initiator to test the bus does not achieve bus saturation or exercise the arbitration to ensure mutually exclusive bus use. A test using two initiators allows the bus occupancy to be increased and the interleaving of transactions from different initiators to be observed as can be seen in the resulting trace shown in Figure 9.2.

To avoid clashes over the same target, in this test the processor fetched instructions from the ROM, and performed data operations on another target (the AMULET3 test port). The instructions were a mixture of moves that fail their condition code check (and are thus discarded, giving a high instruction fetch rate) and load-multiple data instructions which give the fastest cycles on the CPU data port.

Figure 9.2 shows a trace of the (fastest) MARBLE command/address channel cycles taking an average (over the five transfers between the cursors) of 13.6ns each (giving a throughput of 73 million transfers/second). Of the 13.6ns, the channel is actually idle for between 2.6ns and 6ns for the five cycles shown, an average of 3.7ns per cycle.

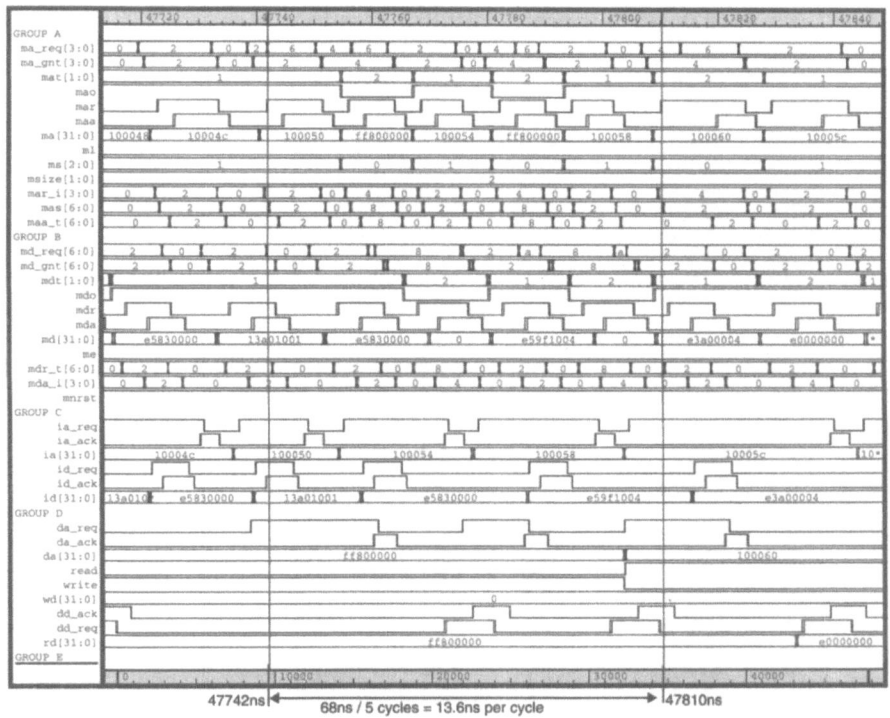

Figure 9.2: Trace showing interleaved CPU fetches and data accesses

9.2.3 Two initiators accessing the same target

Figure 9.3 shows the results of a test with two initiators (in this case both the CPU instruction and data ports) accessing the same target (the ROM).

The test results, as shown in Figure 9.3 show both MARBLE and the ROM accesses running at an average of 56 million transfers/second with the ROM kept busy all of the time. This test runs slightly faster than the single initiator to single target test because the command channel arbitration must be performed sequentially in that test whereas it can be pipelined here.

This confirms that the throughput in the first test was limited by the target performance, and not the initiator. The overlays on Figure 9.3 show the access time of the ROM (from the "ardy" address request to the "donr" data request) and the early arbitration by the ROM target interface before the read data is available from the ROM (visible as the assertion of "md_req[2]" before that of "donr").

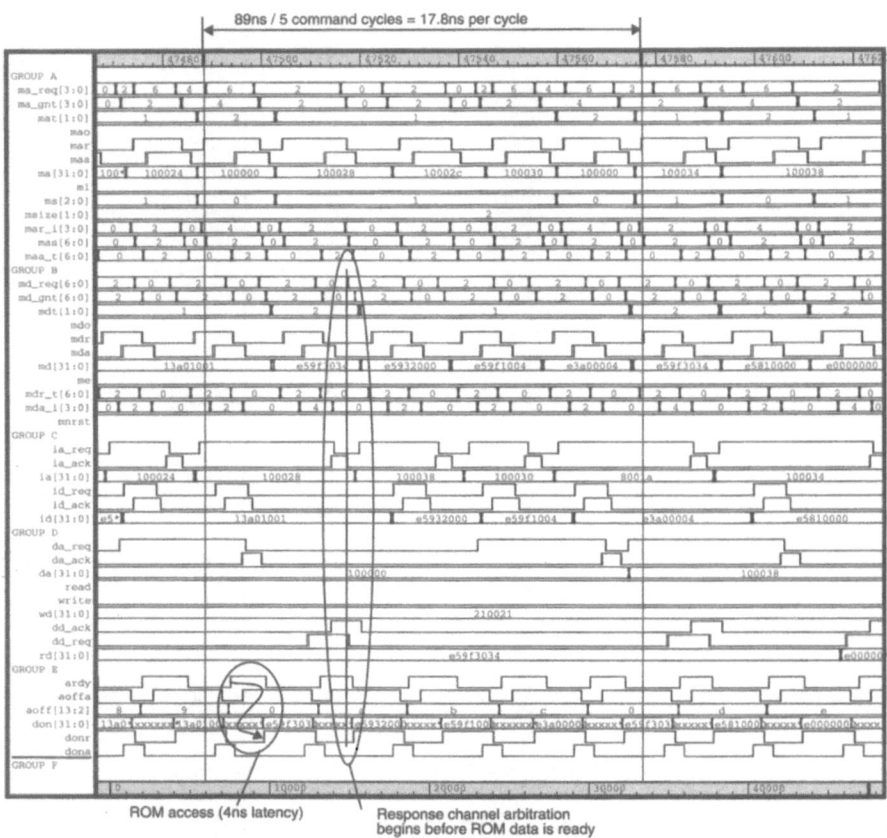

Figure 9.3: CPU fetches and data accesses both using the ROM

9.2.4 Three initiators accessing different targets

None of the earlier tests saturated the bus and so a scenario with three contending initiators was devised. This test involved activity in three initiators and four targets and performed the following actions:

- AMULET3 fetching (and then discarding) instructions from the ROM;
- AMULET3 performing data accesses to another MARBLE target;
- the DMA controller autonomously reading from the SOCB and then writing to the DMA controller programming registers.

Even with this range of activity, the bus is fully occupied for only short durations. This is because the DMA controller takes over 90ns per DMA transfer (one MARBLE read and one MARBLE write), and the processor's bus requirements vary as its internal pipeline fills up with instructions/load dependencies and then drains again, giving a quite complex overall system behaviour.

The longest period of intense bus activity in the simulation was only three cycles. One instance of such activity is shown in the trace in Figure 9.4 where the cycles between the cursors take an average of 12.0ns each (the fastest is about 11.5ns) corresponding to a throughput of 83 million transfers/second. Of this, the signalling lines are idle between transfers for around 2.6ns, although the bus is now fully occupied and this time cannot be reduced or used by other transfers with this implementation for the reasons discussed in Section 9.3.3.

Other behaviours highlighted in Figure 9.4 include:

①a group of sequential fetches with the MS[0] bit set to indicate their sequential relationship even though they do not occur consecutively on the bus;

②the processor waiting (held up by the arbiter) whilst the DMA controller uses the address channel, and then being granted the bus once the processor lowers its arbitration request;

③an example of the pipelined arbitration where the next grant is issued to an initiator whilst the bus channel is still in use by the previous owner;

④an idle period between transfers, even though the bus is heavily loaded by three initiators;

⑤the arbitration latency incurred in starting a new transfer on the bus when arbitration cannot be hidden behind an existing transfer;

⑥the DMA controller reading one byte from the synchronous on-chip bus via the MSB;

⑦an example of the major benefit of supporting split transactions in that other initiators (the CPU fetch and data ports) can use the bus whilst the DMA controller is stalled waiting for read data from the SOCB;

⑧an example of response cycles occurring in a different order from their corresponding command cycles.

9.3 Analysis of delay distribution

Table 9.1 shows the durations of the active phases of some of the fastest handshakes observed on MARBLE (these are for an instruction fetch from the ROM by the AMULET3 core).

These timings include the delays caused by:

- Signal propagation of the transitions on each of the distributed signalling wires. These take between 0.2ns and 0.4ns of each of the durations shown in the table (taking about 1ns in total for the propagation of the four signalling events in a cycle). These times are typical case (the basic delay), not including any crosstalk effects as discussed in Chapter 4.

- Latch controller activity within the target which takes 1.7ns of the MAR+ $\rightarrow$ MAA+ period, and similarly at the initiator 1.7ns of the MDR+ $\rightarrow$ MDA+ period. See Section 9.3.4 for further details of the time taken by the latch controllers.

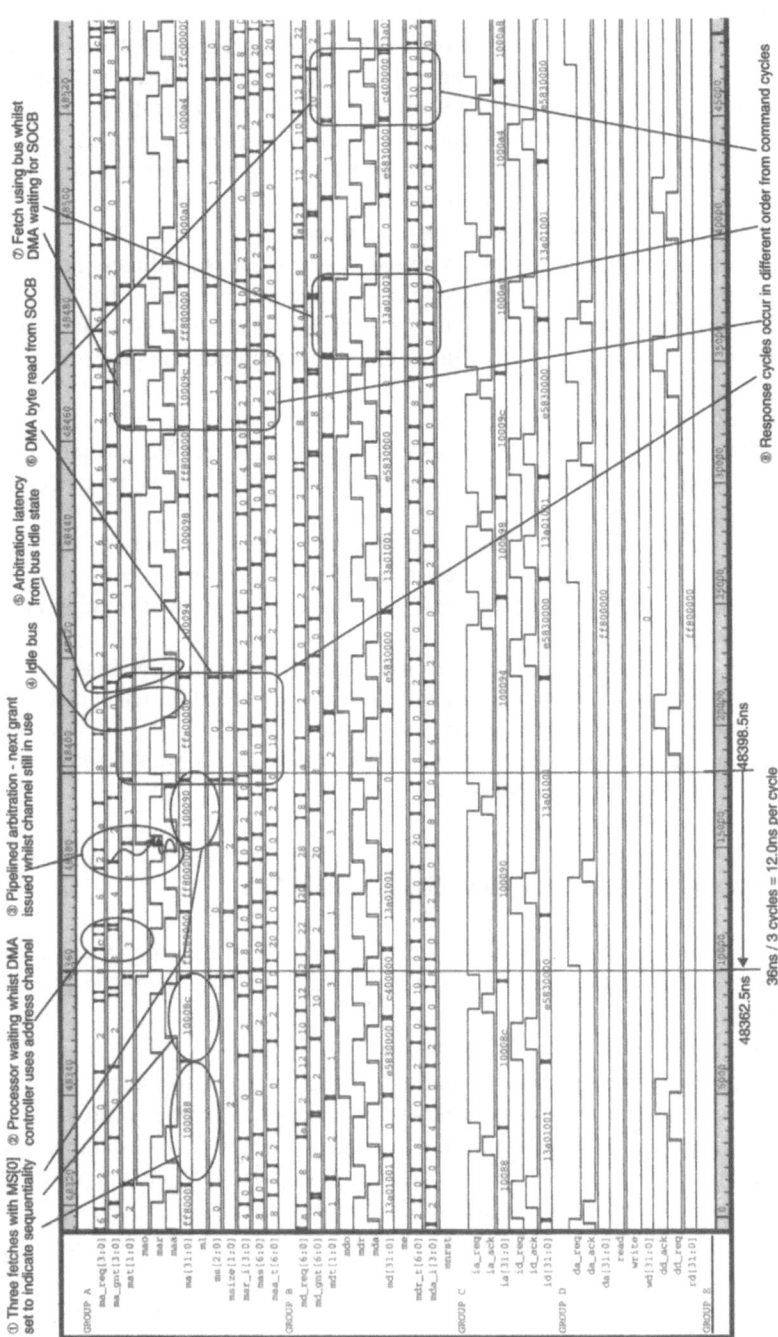

Figure 9.4: Three initiators accessing four targets

Table 9.1: Handshake phase durations for a CPU fetch from ROM (from Figure 9.4)

handshake phase	signalling interval	duration (ns)
command/address setup	MA_GNT+ → MAR+	2.5
command channel push	MAR+ → MAA+	2.9
command channel pull	MAA+ → MAR-	5.0
command channel return-to-zero	MAR- → MAA-	1.8
exception response/read-data setup	MD_GNT+ → MDR+	3.0
response channel push	MDR+ → MDA+	4.3
response channel pull	MDA+ → MDR-	2.9
response channel return-to-zero	MDR- → MDA-	2.4

The remainder of the cycle time is used for address decoding, setup of the bundled signals and activity in the sender (which is tightly coupled to the bus channel) as discussed below.

9.3.1 Centralised and distributed decoding

The command/address channel of MARBLE uses a centralised address decoder to determine which target should respond to a transfer, whilst the response/data channel uses a distributed decoding technique. The decode on each channel takes a similar time, around 0.2ns (less than 2% of the total cycle time), but a direct comparison of the performance of these two mechanisms is not feasible here since the command channel decoder map is more complex than that for the response channel.

9.3.2 Arbitration

Two tree arbiters were used in MARBLE: a 2-level arbiter tree for the command channel, and a 3-level arbiter tree for the response. In each case the uppermost level is a single MUTEX and the other levels use tree arbiter elements as shown in Figure 9.5. The arbitration delay is made up of:

- request propagation: 0.2ns;
- MUTEX propagation (no contention): 0.4ns;
- grant propagation: 0.2ns.

These paths are shown in Figure 9.5. The rise times of the request and grant signals between the arbiter and the device are around 0.2ns each.

Arbitration thereby imposes a latency of about 0.4ns per stage of the arbiter plus 0.4ns for the propagation down the long interconnecting wires, with command channel arbitration taking around 1.2ns and response channel arbitration taking around 1.5ns in total.

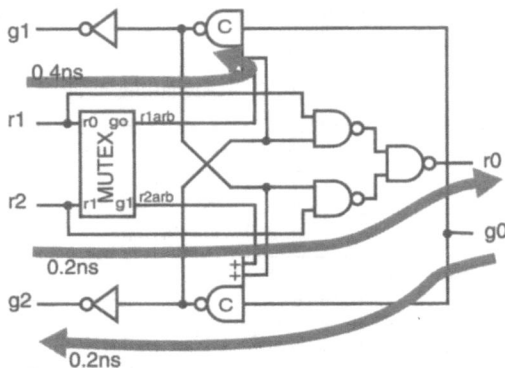

Figure 9.5: Tree arbiter element delays

The pipelining of arbitration for the next cycle on a channel when there is current activity on the channel means that the next initiator is waiting to drive data onto the channel as soon as it is available (indicated by acknowledge low) as intended. Thus arbitration latency has no effect on the bus saturation throughput. This behaviour is visible in some of the earlier traces (and annotated in Figure 9.4) where the assertion of the next grant can clearly be seen to occur before the channel's activity ceases.

9.3.3 Data drive setup time

A breakdown of the "lost time" between cycles on the command channel of the bus is shown in table 9.2 . The results for the response channel are similar with over 1ns absorbed by the delay element in the feedback path used to indicate (to the local controller) when the bus lines have been driven, so that the bus signalling event may be issued.

Table 9.2: Delays in driving the bus channel

Delay period	Duration (ns)
Arbitration grant → drive-enable (channel idle when granted)	0.9
Idle channel → drive enable (granted before bus becomes idle)	0.3
Drive enable signal asserted → data stable on bus channel	1.0
Drive enable signal asserted → data-driven signal asserted	1.4
Data-driven indication → central bus channel request	0.9

From these numbers, there is a total of 0.9ns+(1.4ns-1.0ns)=1.3ns between the bundled data being nominally stable and the corresponding signalling event indicating this to any receivers. This is sufficient to allow for any variations in propagation times due to crosstalk as discussed in Chapter 4.

9.3.4 Pipeline latch controller delays

As discussed in Section 8.2.2, additional buffers were added into the latch controllers to give improved amplification and ensure that the latches remain open for sufficient duration to allow new data to pass through. These delays are necessary to ensure correct functionality, but they increase the forward going latency of the latch and the cycle time, both of which are critical to performance. For the basic normally-closed latch controller (shown in Figure 8.3a) used in the MARBLE designs these delays are as shown in Table 9.3.

Table 9.3: Latch controller performance

	Normally Closed (ns)	Normally Open (ns)
forward latency (input request rising → output request rising)	0.9	0.6
input latching time (input request rising → input acknowledge rising)	1.7	0.9
total input cycle time (input request rising → input acknowledge falling)	2.2	1.5

Table 9.3 also shows the corresponding results for the normally-open latch controller used to store the tag and opcode control information internally at the MARBLE target bridge interface. Even allowing for the difference in loading of the latch controller lines between these latches, the choice of a normally closed latch (for power reasons) has added around 0.3ns to the latency and 0.6ns to the cycle time of every operation on a bus channel. It should, however, be noted that normally open latch controllers tend to the behaviour of their normally closed counterpart if new input data is presented to the latch before the output channel has completed its return-to-zero phase.

The delay introduced by the latch controller is a significant proportion of the entire cycle time of the bus channel (with the fastest cycles actually occupying the channel for 11.0ns, and taking 13.7ns in total).

9.3.5 Sender activity

The MARBLE interfaces used in the AMULET3H were deliberately designed so that they were transparent to the sender on each channel, directly coupling the relevant input channel's protocol to that of the equivalent bus channel. This simplified the design of the control circuits in the interfaces. The consequence of this design choice is that a slow sender with a long delay during the return to zero part of the cycle (ack+ → req-) can hold up the bus. Such behaviour in the local-bus bridges accounts for most of the 5ns between MAA+ and MAR- shown in Table 9.1. The ROM controller which served as the target for the simulation results shown in Table 9.1 was much better in this respect, hence the 2.9ns duration of the MDA+ → MDR- phase of the response/ data cycle.

9.3.6 Performance summary

The performance of a bus is typically characterized by its maximum throughput and the latency incurred by transfers across the bus. The maximum throughput of MARBLE has been measured, as discussed above in Section 9.2.4, as 83 million transfers/second.

The read latency of MARBLE can be obtained by summing the delays incurred in a transfer from the presentation of the command to the initiator interface until the return of a response from the same interface to the initiator. This of course includes the delay of the target device. The full summation may thus be expressed as:

$$\text{Latency} = T_{\text{command transfer}} + T_{\text{target}} + T_{\text{response transfer}}$$

where the delays of the command and response transfers have been measured from the single initiator test as:

$$T_{\text{command transfer}} = 4.0\text{ns} + T_{\text{command arbitration}}$$

and

$$T_{\text{response transfer}} = 8.2\text{ns} + T_{\text{response arbitration}}$$

Thus the total read latency $= 12.2\text{ns} + T_{\text{target}} + T_{\text{command arbitration}} + T_{\text{response arbitration}}$

Which with the 4ns ROM access time, 1.2ns command channel arbitration and hidden response-channel arbitration gives a total ROM read latency of 17.4ns.

9.4 Hardware requirements

The hardware required to implement MARBLE adds little cost to a design. The largest area requirement of a MARBLE bus is the wiring, and the use of a tristate approach keeps this at an acceptable level. The sizes of the controller modules used in the AMULET3H are shown in Table 9.4, based on an average of four transistors per gate. Much of the cost involved in the initiator and target interfaces is due to the pipeline latches, with each interface having around 64 bits of latch on the datapaths of the channels (32-bit address latch and 32-bit write-data latch at the target, 32-bit write buffer and 32-bit read-data latch at the initiator).

Table 9.4: MARBLE interface and bus control hardware costs

Module	No. of gates
Initiator interface	570
Target interface	600
4-way tree arbiter	40
7- or 8- way tree arbiter	100
Address decoder	90

9.5 Comparison with synchronous alternatives

Synchronous buses, as introduced in Chapter 3, use a clock to regulate the transfer of data between devices. Making a direct performance comparison between on-chip buses is difficult since their performance is restricted by the particular process technology used in a given implementation. Furthermore, the bus clock used in a design may not be the maximum possible frequency that the bus could use, but instead a convenient frequency that can easily be derived from the main processor clock (e.g. half its frequency). One example of the flexibility of an asynchronous on-chip bus is that its performance need not be constrained in this way. However, Table 9.5 shows a comparison of the throughput and read-latencies for MARBLE (in 0.35 micron CMOS) with those for PI-Bus, AMBA-AHB and CoreConnect based on clock-frequencies published in data sheets and product specifications. Note that latency calculations have been performed assuming that the bus was previously idle, with the default grant (if one exists) being given to the contending initiator.

Table 9.5: Bus performance figures

Bus	Data-path width (bits)	Frequency (MHz)	Throughput (MB/s)	Mimimum bus imposed read latency from idle (ns)	CMOS feature size (μm)	Number and type of metallisation layers	Product / Datasheet
MARBLE	32	83	332	14	0.35	3 Al	AMULET3H
OMI PI_Bus	32	50	200	40			Specification v0.3d
AMBA ASB	32	18	72	110	0.6	3 Al	Cirrus Logic CL7110
AMBA AHB	32	150	600	14	<0.25		Predicted for ARM10
CoreConnect OPB	32	50	200	40	0.25	5 Cu	PowerPC 405GP
CoreConnect OPB	32	66	264	30	0.18	Cu	PowerPC 440 Core
CoreConnect PLB	64	100	800	20	0.25	5 Cu	PowerPC 405GP
CoreConnect PLB	128	133	2128	15	0.18	Cu	PowerPC 440 Core

This table shows MARBLE achieving a throughput between that of the high-performance AMBA-AHB and CoreConnect PLB (both of which use a gate-multiplexed approach instead of tristate bus lines) and the mid-range CoreConnect OPB.

However, the bus imposed read latency of MARBLE is comparable to that obtained with synchronous buses running at around 150MHz. This is a consequence of every transfer on a synchronous bus with overlapped address and data phases requiring a minimum of two clock cycles, one for the address transfer and then one for the data transfer. Furthermore, arbitration on a synchronous bus by a device other than that with the default grant would require a further clock cycle giving bus-imposed read latencies of the order of 20ns for the 150MHz buses. (Note that the figure quoted for MARBLE includes command channel arbitration delays since there is no default grant mechanism.)

10. Conclusion

Future SoC devices will require high performance on-chip buses. These are necessary to support design methodologies based upon component reuse and the growing market in SoC intellectual property. The work described in this book has identified and presented asynchronous solutions to the problems that must be overcome in the design of an on-chip system bus. The work has resulted in the specification of an asynchronous bus architecture for use in such systems.

The resulting MARBLE bus architecture supports many of the features found in synchronous on-chip buses and in both synchronous and asynchronous off-chip buses, providing a similar throughput and imposing a similar latency penalty as for equivalent synchronous buses. Asynchronous implementations of common bus features such as pipelined transfers, overlapped arbitration, atomic transactions and split transactions have all been demonstrated by MARBLE along with support for precise exceptions. In AMULET3H the MARBLE architecture has been used in a SoC design, in its simplest form with only one outstanding command allowed per initiator.

MARBLE has been a key enabling factor in the AMULET3H design process, allowing separate work on each of the major macrocells. These were designed using vastly different techniques, ranging from the full-custom layout of the processor to the synthesis of the DMA controller from a high-level channel-based programming language. The major advantage of MARBLE over its synchronous counterparts is that it allows the benefits provided by asynchronous design, as introduced in Chapter 1, to be extended throughout the system, not constrained to a single macrocell. It also offers a clearly defined solution for the globally asynchronous interconnection of many macrocells which may operate locally under any timing model, be it asynchronous or synchronous.

10.1 Advantages and disadvantages of MARBLE

Many arguments are proffered both for and against asynchronous circuits. Those relevant to the use of asynchronous techniques for SoC interconnect as addressed in this book, demonstrated by the MARBLE bus architecture, and realised in the AMULET3H system are examined below.

10.1.1 Increased modularity

One of the principal benefits often quoted by advocates of asynchronous design is improved modularity when compared to a synchronous alternative. This is one of the strongest arguments in favour of using an asynchronous approach to SoC interconnect. The use of MARBLE in the AMULET3H system clearly illustrates this advantage, allowing each macrocell to be designed independently to perform to the best of its abilities without being constrained to have to fit within a given clock period.

For example, a ROM access takes 4ns, whereas an SOCB read access takes a minimum of 22ns (1.5 clock cycles at 66MHz). The DMA controller and the processor can use either of these peripherals, or the others in the system without the imposition of any additional delay, unlike in a synchronous system where the delay in an access to a device would have to be constrained to be a multiple of a fixed clock period. The asynchronous SoC interconnect approach thus provides a greater modularity than the synchronous alternatives by avoiding the problems of clock distribution.

The other implication of this modularity is the ease with which performance improvements may be made. For example, in future designs the processor may be further optimised to gain, say, another 20% performance increase and the ROM access time may be reduced or other faster targets such as an SDRAM interface may be added. The MARBLE bus will be able to support such changes without modification, and existing peripherals, including the DMA controller, will still function correctly unmodified. To perform the same improvements in a synchronous system would require either an increase in the global clock frequency (requiring every device to be re-examined and possibly modified to ensure it meets the new timing constraints) or the introduction of multiple clock domains which brings additional circuit complexity.

10.1.2 Avoidance of clock-skew

The asynchronous sections of the AMULET3H system show how complex systems can be designed in the absence of a clock, thus avoiding the problem of clock skew altogether. Where a clock is necessary, its use can be localised to a sufficiently small region that clock skew is easily managed, with asynchronous techniques used for global system interconnections. AMULET3H does just this, using MARBLE as the global system interconnect with some peripherals implemented synchronously, although in this chip the silicon areas occupied by circuits designed using the two different timing methodologies are similar.

10.1.3 Low power consumption

The system bus is only one part of the entire design, but it nonetheless consumes power when performing transfers. When heavily loaded, there would be little difference between the synchronous and asynchronous approaches in terms of their power consumption. However, the use of asynchronous techniques for the system bus offers real benefits when the bus is lightly loaded since it has zero quiescent power consumption and can switch from idle to full throughput immediately with no added complexity. Of course, to design a completely asynchronous system an asynchronous bus is required, and so MARBLE is also an enabling factor for the design of large, clock-free systems.

10.1.4 Improved electro-magnetic compatibility (EMC)

No EMC measurements or simulations have yet been performed on the AMULET3H design, although the results are expected to be similar to those obtained with the AMULET2e microprocessor where the radiated energy was shown to be of lower amplitude and less concentrated into narrow spectral bands than its synchronous counterpart. Again one of the more significant advantages offered by MARBLE in this area is the ability to construct large completely asynchronous systems without the use of a clocked macrocell interconnect. It is thus an enabling factor in the application of asynchronous techniques for reducing EMC throughout the system, as the use of a synchronous bus would subvert many of the improvements obtained by using a large proportion of asynchronous macrocells.

10.1.5 Performance

MARBLE's performance is similar to that of its synchronous counterparts, although the read latency is marginally better than synchronous buses capable of the same throughput. This comes from not having to delay the response transfer to start on a clock transition, allowing it to start as soon as the target is ready. From the work presented in this book it would appear that there is little difference between the performance of synchronous and asynchronous SoC interconnect buses. However, as feature sizes shrink and crosstalk worsens, asynchronous approaches may offer an advantage through their ability to give average case performance instead of the worst-case performance offered by synchronous designs.

10.1.6 Risk of deadlock

MARBLE and the AMULET3H design have been extensively simulated without the detection of any deadlocks due to the bus protocol or the bus interface circuit implementations. However, neither the bus protocol nor the circuit designs have been completely formally verified. The bus protocol has been partially investigated using formal techniques by the Formal Methods Group at the University of Manchester, but this was performed more as a test of the Rainbow toolkit [13] than as a verification of MARBLE. A full formal verification would be of significant value in proving the robustness of the protocol.

10.1.7 Timing verification

Timing analysis of the MARBLE bus is substantially more difficult than for an equivalent synchronous bus. For the correct operation of a synchronous bus, a one-sided timing constraint must be met:

For an asynchronous bus such as MARBLE the constraint must also take into account the effect of delays and crosstalk on the signalling lines, giving a two sided constraint:

$$\begin{matrix} \text{data} \\ \text{propagation} \\ \text{delay} \end{matrix} \quad + \quad \begin{matrix} \text{worst case} \\ \text{additional} \\ \text{crosstalk} \\ \text{delay} \end{matrix} \quad < \quad \begin{matrix} \text{signalling} \\ \text{propagation} \\ \text{delay} \end{matrix} \quad - \quad \begin{matrix} \text{best case} \\ \text{crosstalk} \\ \text{speed-up} \end{matrix}$$

Such constraints are difficult to check statically because of the crosstalk dependencies, and are not easy to check dynamically. Tools were not available for making these checks, and so substantial margins based on the results of the crosstalk simulations shown in Chapter 4 were built into the senders on both the command and response channels to ensure correct timing operation under all circumstances.

10.1.8 Design complexity

MARBLE uses the single rail bundled data design style. As a result, the wire and gate requirements of a MARBLE bus are similar to those of an equivalent synchronous bus. As with many asynchronous circuits, the controllers may be a little more complex than a synchronous equivalent due to the request-acknowledge signalling, but this comparison is difficult as circuit details are not readily available for the proprietary implementations of equivalent bus interfaces for synchronous buses such as AMBA and CoreConnect.

10.1.9 Reliable arbitration

One of the lesser cited benefits of asynchronous design techniques is the high reliability of the arbitration of asynchronous inputs. This means that a bus such as MARBLE will never give incorrect arbitration operation, although there is a small probability that the arbitration may require an infinite time.

Of course, a synchronous arbiter in a synchronous circuit with only one clock domain will also never suffer arbitration failure when arbitrating between internally derived signals. However, when a circuit contains multiple timing domains, possibly including a mixture of synchronous and asynchronous subsystems, there is a real risk of arbitration failure if the arbitration is performed using synchronous techniques. Waiting for a second, or third clock period can reduce the failure rate, but further increases the arbitration latency.

An asynchronous arbiter does not suffer from this problem because the arbitration is not constrained to fit in a fixed time window (of a defined number of clock periods). Bus arbitration can also be much faster in an asynchronous SoC bus than in a synchronous one, as in most cases there will be no contention on the arbiter at the time when it resolves the ordering of the requests. Even when there is contention, there is a high likelihood that the metastability will only last for a short duration (typically much less than the clock period of equivalent synchronous buses). In MARBLE this means that typically the arbitration latency is of the order of 1.5ns. This low latency means that split transactions can be implemented on top of a decoupled bus protocol for EVERY transaction at negligible cost in terms of performance or hardware.

10.2 Improving the MARBLE bus

MARBLE achieves a throughput better than the mid-range SoC buses and is not far behind the higher performance SoC buses developed for use at frequencies above 100MHz, allowing for implementation technology differences.

The implementation of MARBLE used in the AMULET3H was designed conservatively to prove the viability of asynchronous SoC interconnect. As such it can be improved in a number of ways including the following.

10.2.1 Separating the read- and write- data paths

MARBLE passes the write data over the response channel using the same tristate lines used to pass the read data. This approach keeps the wire resource requirements of MARBLE low, but limits the performance and the support for precise exceptions. Adding separate datapath wires as part of the command channel to carry write data (as per mapping d in Figure 6.2 on page 64) would mean that:

- the target would not have to wait for a few nanoseconds after receiving the command whilst the write data is pulled across the response channel;
- precise exceptions could be bridged to another multimaster bus in all circumstances.

10.2.2 Less conservative drive overlap prevention

MARBLE used conservative timing assumptions regarding the switch-on and switch-off of the tristate drivers to ensure that not only was there no overlap, but that there was a significant period (usually at least one phase of a handshake) between one driver switching off and another switching on. Simulations show that the drivers actually switch fairly quickly (although sufficient time must be allowed to accommodate variations due to crosstalk effects) and sufficient margin to avoid significant drive overlaps would still be obtained by allowing:

- new commands to be driven when request is low (as opposed to when both request and acknowledge are low);
- new responses to be driven when request is low (as opposed to when both request and acknowledge are low).

The latter is definitely safe for a "write-with-command" scheme, and should also be safe for a "write-with-response" scheme such as that of MARBLE where there is also the possibility of a clash between pulled write-data and subsequent pushed read-data drive.

10.2.3 Allowing multiple outstanding commands

The AMULET3H implementation of MARBLE permits only one outstanding command per initiator. Consequently a single initiator cannot saturate the bus as there will always be a delay between the start of the response cycle and the start of the subsequent command cycle whilst the initiator arbitrates for the command channel and drives the command onto it. Allowing an initiator to have multiple outstanding commands would permit it to start the second command cycle immediately after the first. The number of outstanding commands required to allow saturation of the bus depends on the speed of the target devices, but four should probably be sufficient; this would mean that a response must be received less than four command cycle durations after the start of the first command to allow the fifth command to be issued without delay after the fourth command. Four outstanding commands should also satisfy any pipelining needs with most target peripherals, allowing one command and one response to be in transit on the bus with two other transfers active within the target device. Of course MARBLE could be extended to support many more than four outstanding commands if required.

10.3 Alternative interconnect solutions and future work

Crosstalk effects, as discussed in Chapter 4, mean that MARBLE has to include significant delays in its signalling paths to ensure that the bundling constraint is met. To minimise the size of these effects, the wires were spaced at a greater separation than the minimum allowed by the design rules. The alternative interconnect solutions discussed below may provide a better performance at similar cost.

10.3.1 Changing the interconnect topology

A shared bus suffers from large crosstalk effects because the connecting wires must be long so as to connect all of the devices together. The effects of crosstalk could be reduced by changing topology from a bus to a centralised hub approach with dedicated unidirectional point to point connections between the hub and each device. This would greatly reduce the effects of crosstalk through shortening the lines, and allowing extra amplification repeaters to be used. Higher cycle-rates would thus be possible.

Alternatively, for a lower resource cost, a ring network as illustrated by Figure 10.1 could be used to connect the devices in a loop using short unidirectional point to point channels which again could include signal repeaters if necessary.

This topology would allow behaviour similar to that of a bus in that a command could be flowing from initiator to target, whilst a response was returning around the other part of the loop. Furthermore, a greater degree of concurrency is supported by such an architecture than when using a bus. If the performance of a single ring system is insufficient, two rings could be used, one for commands and one for responses, possibly flowing in opposite directions.

As feature sizes shrink and wire delays and crosstalk effects become more significant, it may be that such approaches can offer much better performance than a shared bus at a similar cost.

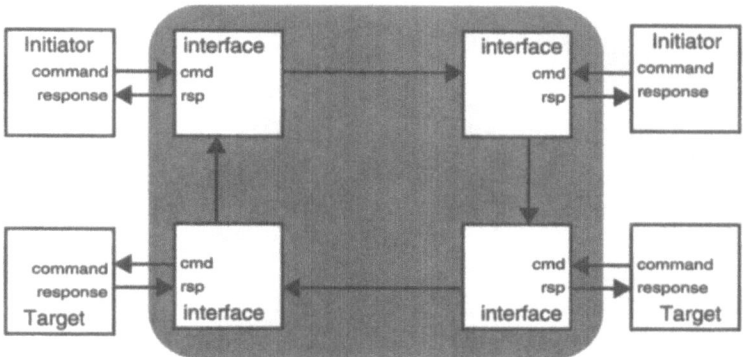

Figure 10.1: A ring network

10.3.2 Changing to a delay-insensitive data encoding

Interestingly, crosstalk does not cause a slow-down of every transfer; its effects are dependent on the data being transmitted, so the worst case slowdown may not occur very often. Using a delay insensitive data encoding would allow these better-than-worst-case scenarios to be exploited. However, again, the penalty may be an increase in resource requirements for the application of such techniques to a SoC bus.

As an example, consider a 1-of-4 one-hot encoding where a symbol is transmitted on one of a group of four wires. Two bits of information are thus conveyed over four wires, giving the same resource requirements as a dual-rail encoding. Within the group of wires, the crosstalk slowdown will be the same.

Any two adjacent wires in the single-rail MARBLE design could be changing level in opposite directions, leading to a significant slowdown. With the 1-of-4 encoding, if the wires are laid out as shown in Figure 10.2, then the worst case should only occur very infrequently since most of the time non-adjacent wires would be switching. Furthermore, the wires could thus be spaced differently, with reduced lateral spacing within a group and a larger lateral separation between groups to trade area for further crosstalk improvements.

Such a delay insensitive code is not easily implemented with a bus type interconnection because of the bidirectional tristate nature of a bus. However, this type of approach would be ideally suited to the alternative interconnect topologies discussed above in Section 10.3.1

Transition signalling
A delay insensitive encoding scheme such as a 1-of-4 one-hot code can be used in conjunction with transition signalling to offer a higher throughput when compared with a return-to-zero signalling scheme. Further investigation would be required to determine if the cost of converting from a level-based signalling scheme to transition signalling over the long bus wires, and then back again at the far end, is worthwhile, but with careful design of the converters it should in theory be possible almost to double the throughput over the same wires.

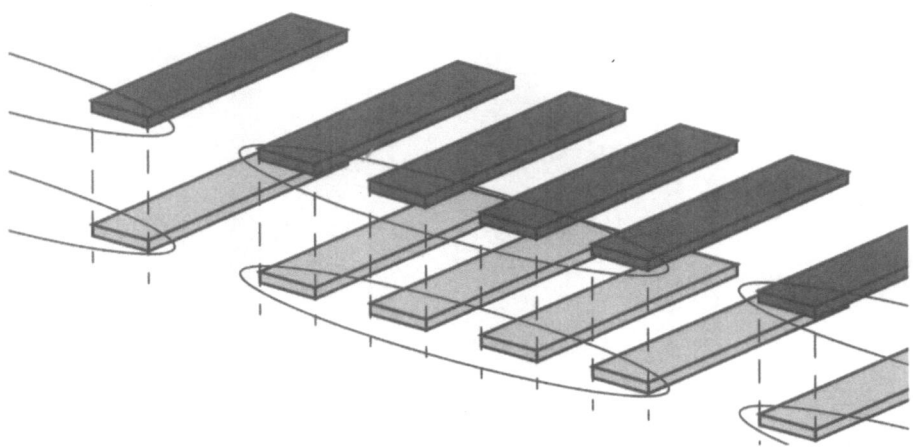

Figure 10.2: A suitable layout for a 1-of-4 encoding

More complex encodings

Other more complex codes, such as a 3-of-6 or a 2-of-7 code, convey more bits per wire than the simpler dual-rail or one-hot codes. For example, a 2-of-7 encoding can represent up to 21 different symbols, sufficient to carry four bits of information. This encoding requires only seven wires whereas a dual-rail or 1-of-4 coded system would require eigth wires to carry the same information. Further investigation would be required to determine if the conversion between conventional single-rail encoding and such complex encodings could be performed sufficiently fast for on-chip communications.

10.4 The future of asynchronous SoC interconnect?

The work in this book has shown the feasibility of asynchronous SoC interconnect with a concrete example to rival synchronous SoC buses in the form of the MARBLE bus. The EMC and modularity advantages of asynchronous design and operation can thereby be extended throughout the system as a whole.

MARBLE has also highlighted the difficulty of using the bundled-data approach for long bus wires, where the crosstalk problem becomes difficult to manage, and in practice this requires manual layout of the bus wires. Whilst this approach is not a problem today, automated routing is becoming more common, and indeed necessary, to meet time-to-market requirements for large SoC designs. In this respect, the synchronous buses and the bundled-data solution used by MARBLE both suffer from the need for extensive timing analysis to be performed.

It is hoped that based upon this book, SoC interconnect is an area where asynchronous techniques can gain a foothold since they offer a clear modularity advantage. However, in the longer term, shared buses will have significant problems delivering the required performance and a delay-insensitive asynchronous interconnect approach using unidirectional point-to-point connections may be a better solution.

Appendix A: MARBLE Schematics

This appendix contains a complete schematic dump for the MARBLE bus interfaces as used in the AMULET3H telecommunications controller chip. These interfaces do not support transfer deferral.

A1   Bus interface top level schematics

The top level schematics of the initiator interface (schematic A1) and the target interface (schematic A2) are structured as shown earlier in Figure 8.2 on page 89.

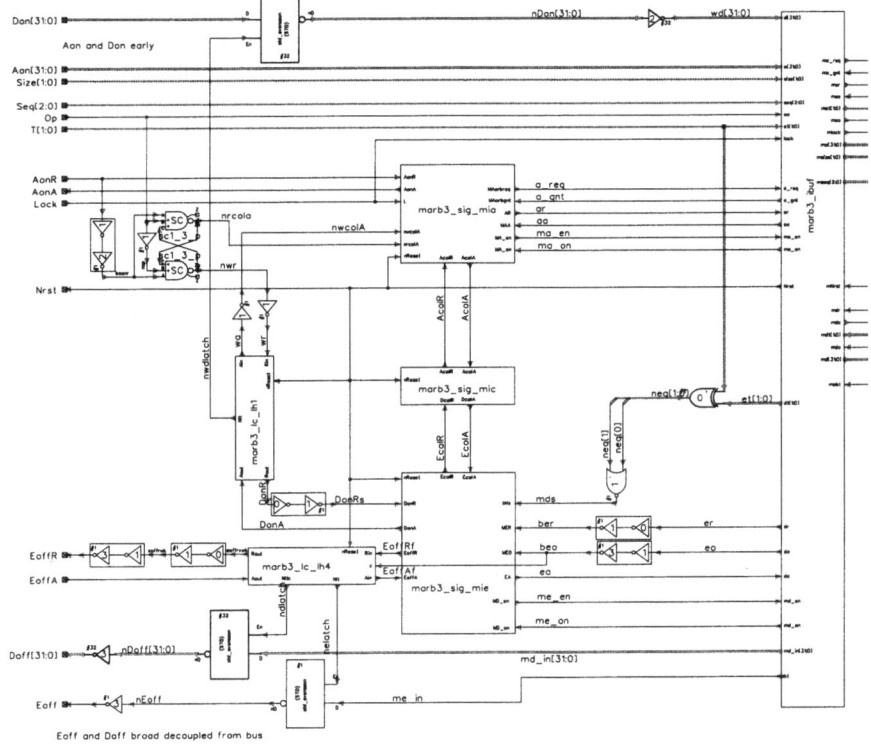

Schematic A1: Initiator interface top-level schematic

A2   Initiator interface controllers

The two main controllers in the initiator are both based upon the multipoint channel controllers introduced in Chapter 5. The command channel controller shown in schematic A3 is a channel initiator capable of locking the multipoint channel, with two input channels, one for the command and one for the token enabling the transfer to be

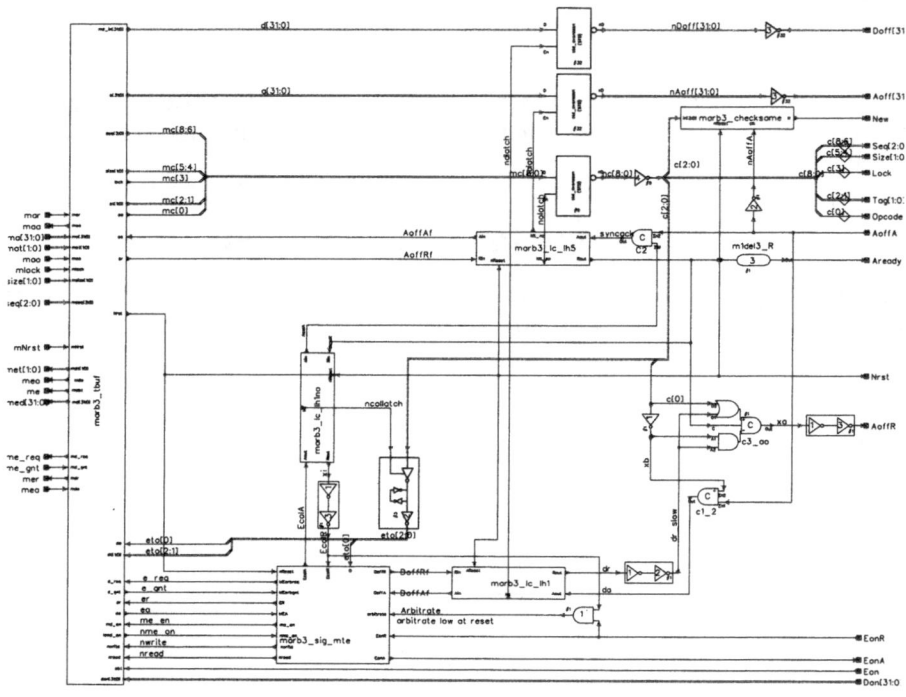

Schematic A2: Target interface top-level schematic

passed out onto the bus. The response channel controller shown in schematic A4 is a channel target capable of receiving the response and either receiving pushed read data from the channel or delivering pulled write data onto the channel depending on the transfer direction as indicated by the incoming MDO line. The throttle for controlling the token flow and issuing the first token into the loop is a simple decoupling circuit derived from a latch controller, with its output primed at reset. Its construction is shown in schematic A5 .

A3 Target interface controllers

The MARBLE interfaces for the AMULET3H system do not support deferred transfers. As a consequence, there is no need for a "command channel controller" at the target, since every incoming command will be accepted. The response controller shown in schematic A6 is based upon the multipoint channel initiator in Chapter 5. It waits until it has been granted the channel, and the response is available, before turning on the drivers and then signalling the request. Any write data is delivered to the target peripheral later in the same handshake.

The logic for checking the initiator id to ensure that the MS[2:0] signals can be filtered correctly to indicate a new, non-sequential transfer when the active initiator is not the same one as for the previous command is shown in schematic A7.

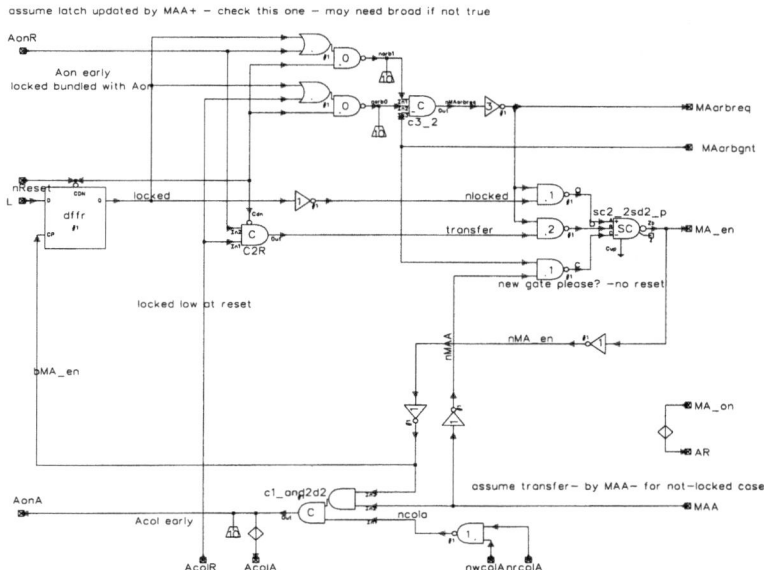

Schematic A3: Initiator command controller

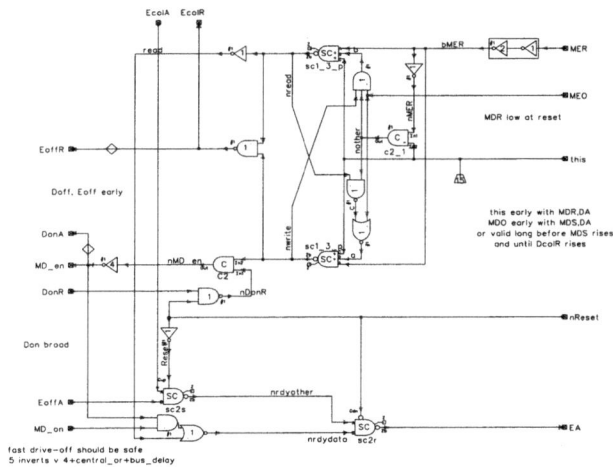

Schematic A4: Initiator response controller

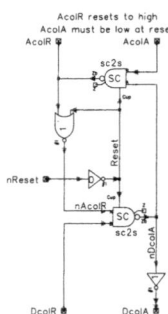

Schematic A5: Initiator token unit

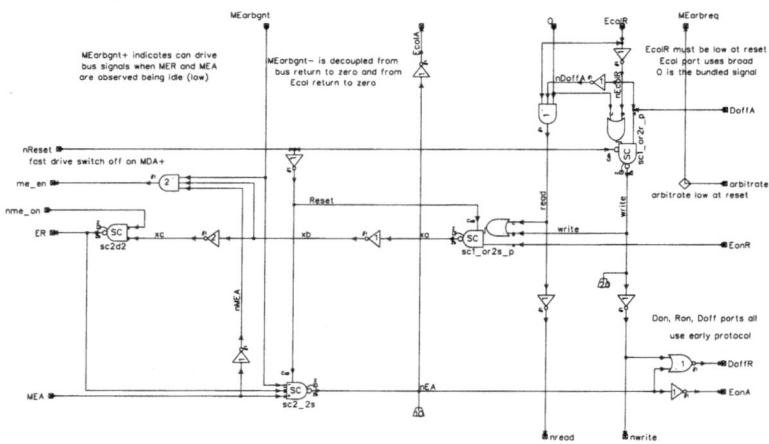

Schematic A6: Target response controller

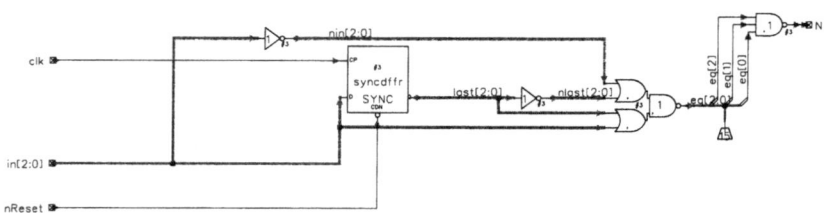

Schematic A7: Target "same-initiator" check

A4 Bus drivers and buffers

Schematics A8 and A9 show the buffers and tristate drivers used to connect signals onto the shared bus wires. The delays to provide sufficient margin to meet the bundling constraint (and allow for crosstalk effects) are included in these schematics.

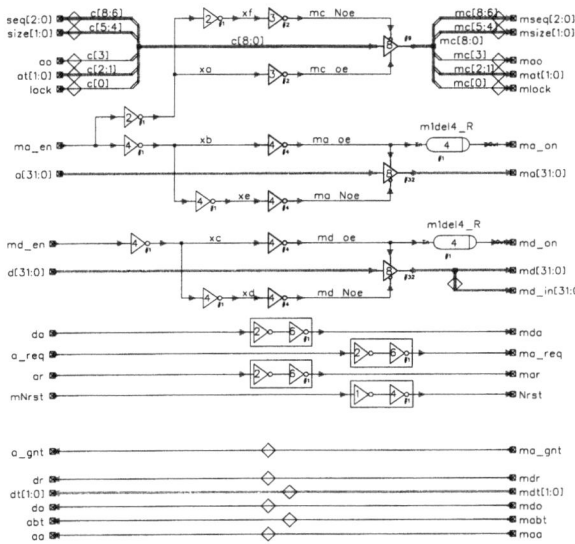

Schematic A8: Initiator bus driver/buffer

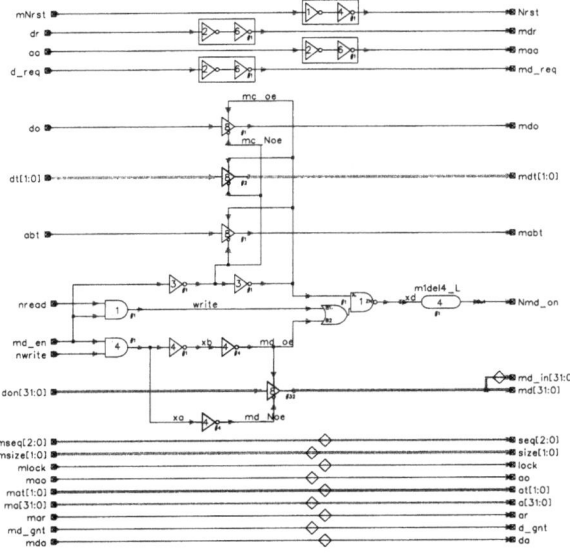

Schematic A9: Target bus driver/buffer

A5 Latch controllers

The four types of latch controller used in the MARBLE interfaces, as described in Section 8.2.2 are shown in schematics A10, A11, A12 and A13.

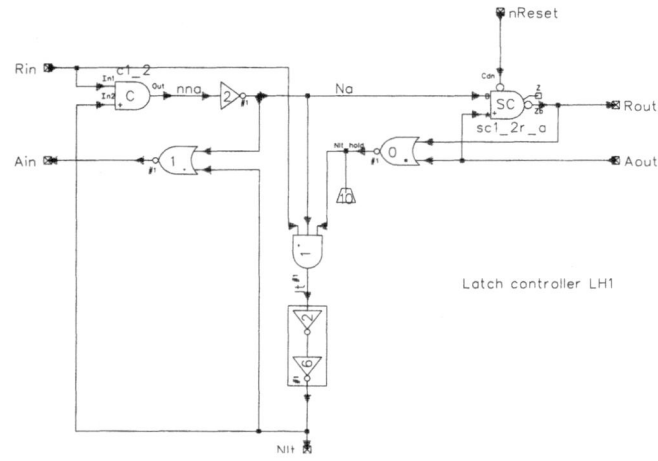

Schematic A10: Normally closed latch controller

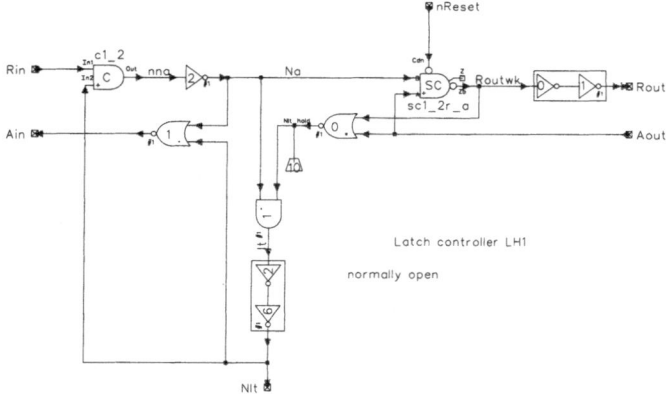

Schematic A11: Normally open latch controller

A6 Centralised bus control units

Schematic A14 shows the construction of the 4-way and 7-way tree arbiters used for the command channel and response channel arbitration. The construction of the tree-arbiter element is shown in schematic A15.

The OR functions that combine the individual requests and acknowledges from the interfaces to form the MAR, MAA, MDR and MDA signals are shown in schematic A16. The amap unit in that schematic consists of an address decoder for the

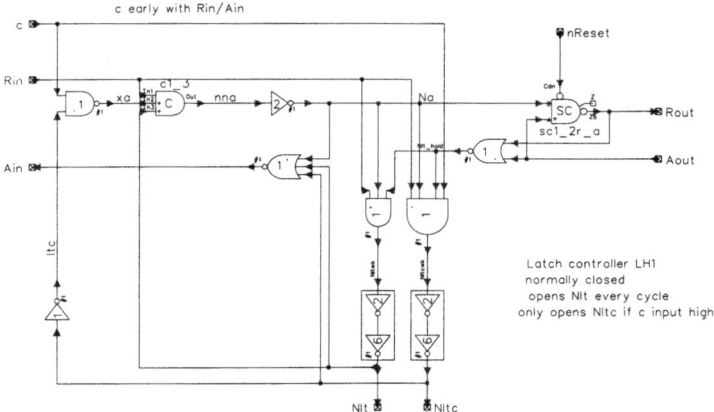

Schematic A12: Conditional open latch controller

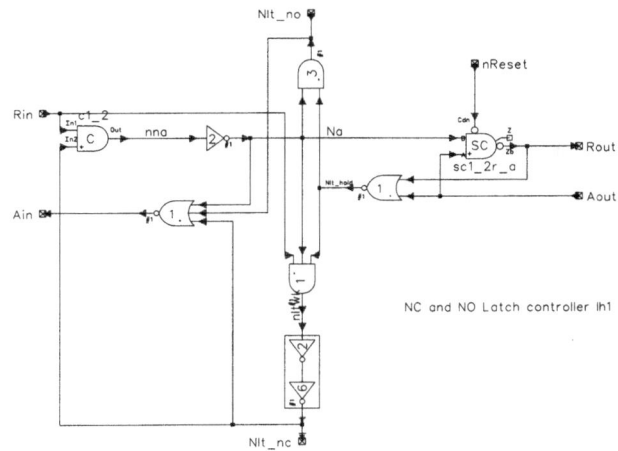

Schematic A13: Normally closed and normally open hybrid latch controller

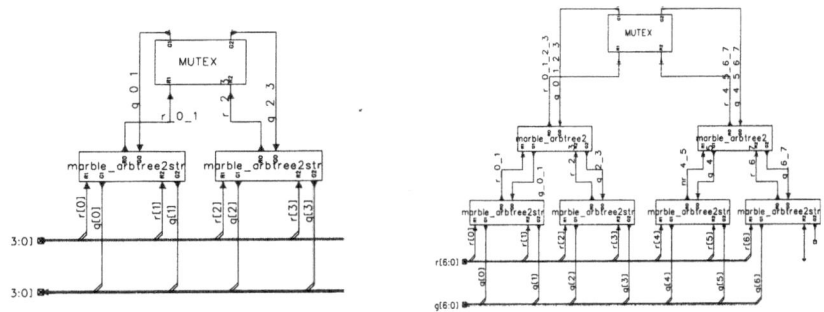

Schematic A14: 4-way and 7-way tree arbiters

Tree Arbiter Element
actually arbitrated call with arbitration in parallel with R0+ −) G0+
Timing assumption: r1arb falls before G0 goes high due to r2arb and vica versa

Schematic A15: Two input tree-arbiter element

AMULET3H address map. This unit, shown in schematic A17 was constructed by Dr S. Temple who specified the address map for the AMULET3H. A simple 8-way decoder was used in the MARBLE testbed described in Section 9.1. The final piece of the centralised bus control logic is the weak feedback bus keepers shown in schematic A18 that prevent the bus lines floating at voltages near the logic threshold of the gates whose inputs they drive.

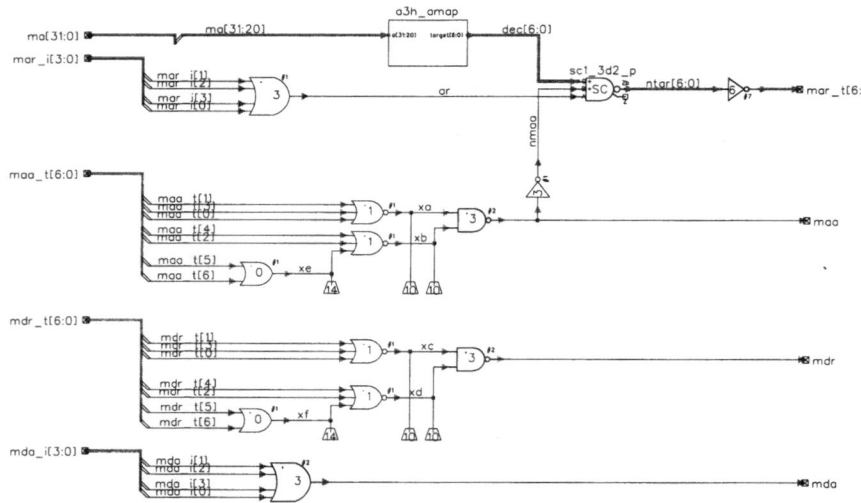

Schematic A16: Centralised signalling OR functions and address decoder

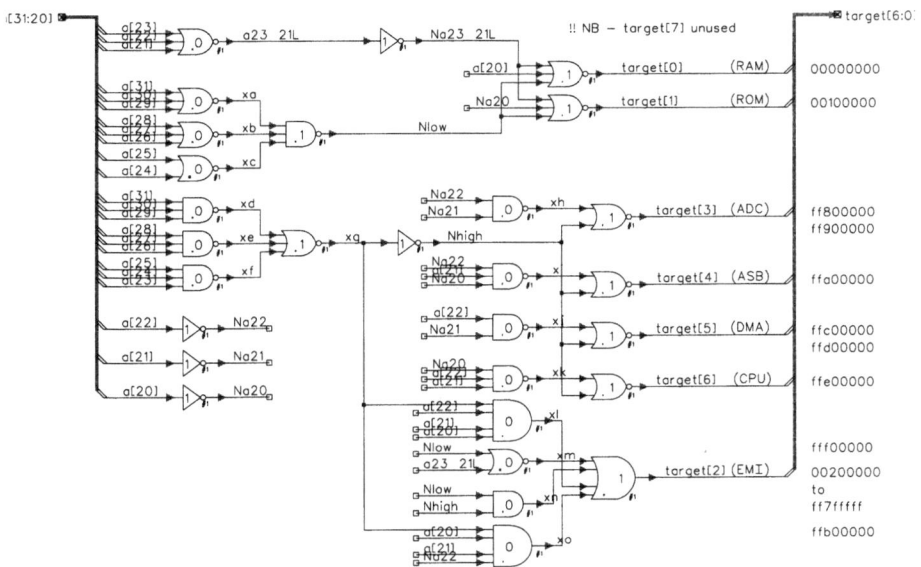

Schematic A17: Address map specific section of address decoder

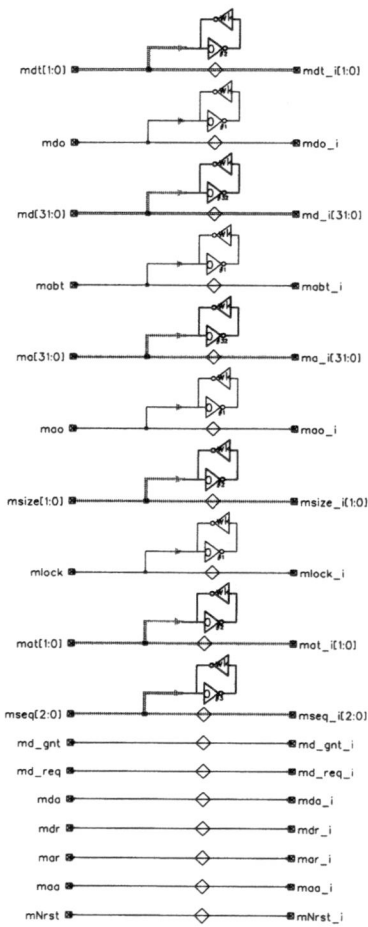

Schematic A18: MARBLE bus keepers

References

[1]   Alpha 21264 Microprocessor Hardware Reference Manual. Compaq Computer Corporation, (July 1999).

[2]   AMBA, Advanced Microcontroller Bus Architecture Specification, Rev D, Advanced RISC Machines Ltd (ARM), (April 1997).

[3]   AMBA, Advanced Microcontroller Bus Architecture Specification, Rev 2.0, Advanced RISC Machines Ltd (ARM), (May 1999).

[4]   AMULET3H - 32-bit Integrated Asynchronous Microprocessor Subsystem. AMULET Group, University of Manchester, UK. version 0.8, 1999. URL http://www.cs.man.ac.uk/amulet/

[5]   Bainbridge, WJ, Furber, SB, Asynchronous Macrocell Interconnect using MARBLE, In Fourth International Symposium on Advanced Research in Asynchronous Circuits and Systems, ASYNC'98. Department of Computer Science, IEEE Computer Society Press, (April 1998).

[6]   Bainbridge, WJ, Furber, SB, MARBLE: An asynchronous on-chip macrocell bus. Microprocessors and Microsystems, 24(4), pp 213-222 (August 2000).

[7]   Bainbridge, WJ, Furber, SB, Delay-Insensitive System-on-Chip Interconnect using 1-of-4 Data Encoding. In Seventh International Symposium on Asynchronous Circuits and Systems, ASYNC2001, IEEE Computer Society Press, (March 2001).

[8]   Bainbridge, WJ, MARBLE: A proposed asynchronous system level bus. In Proceedings of the Second UK Asynchronous Forum, (June 1997).

[9]   Bainbridge, WJ, Bridging between MARBLE and a clocked peripheral bus. In Proceedings of the Fourth UK Asynchronous Forum (July 1998).

[10] Bainbridge, WJ, Multi-way arbitration. In Proceedings of the Fifth UK Asynchronous Forum, (July 1999).

[11] Bainbridge, WJ, Crosstalk analysis for AMULET3/MARBLE. In Proceedings of the Fifth UK Asynchronous Forum, (July 1999).

[12] Bardsley, A, Balsa: An Asynchronous Circuit Synthesis System, Master's thesis (1998), The University of Manchester, UK.

[13] Barringer, H, Fellows, D, Gough, G, Jinks, P, Williams, A, Multi-View Design of Asynchronous Micropipeline Systems Using Rainbow. In IFIP International Conference on Very Large Scale Integration, VLSI'97, 1997.

[14] Bruin, C, Cunningham, F, Esprit project 23031 - ATOM, Cheshire Henbury, (1999).

[15] Chaney, TJ, Molnar CE., Anomolous behavior of synchronizer and arbiter circuits. IEEE Transactions on Computers 22(4), pp 421-422 (April 1973).

[16] Chapiro, DM, Reliable high-speed arbitration and synchronization. IEEE Transactions on Computers C-36(10), pp 1251-1255 (February 1987).

[17] Chu, T-A, Synthesis of Self-timed VLSI Circuits from Graph-theoretic Specifications. Ph.D. thesis, MIT, (June 1987).

[18] Chu, T-A, On the Models for Designing VLSI Asynchronous Digital Circuits, Integration, the VLSI Journal, 4(2), pp 99-113 (June 1986).

[19] Coates, WS, Davis, AL, Stevens, KS, Automatic Synthesis of Fast Compact Self-Timed Control Circuits. In IFIP Working Conference on Asynchronous Design Methodologies, pp 193-208 (April 1993).

[20] Compass Design Automation Tools v8r4.10.0 from Avant!. See http://www.avanticorp.com/compass

[21] Cortadella, J, Kishinevsky, M, Kondratyev, A, Lavagnao, L, Yakovlev, A, Petrify: a tool for manipulating concurrent specifications and synthesis of asynchronous controllers. IEICE Transactions on Information and Systems. E80-D(3), pp 315-325 (March 1997).

[22] Cortadella, J, Kishinevsky, M, Kondratyev, A, Lavagno, L, Yakovlev, A, Synthesis of control circuits from STG specifications. In: handouts of the Summer School on Asynchronous Circuit Design, Technical University of Denmark, (August 1997)

[23] Couranz, GR, Wann, DF, Theoretical and experimental behaviour of synchronizers operating in the metastable region. IEEE Transactions on Computers 24(6), pp 604-616 (June 1975).

[24] Craft, DJ, Improved CMOS Core Interconnect Approach for Advanced SoC Applications, In IP99 Europe, pp 233-246 (November 1999).

[25] Dally, WJ, Poulton, JW, Digital Systems Engineering. Cambridge University Press (1998), ISBN: 0-521-59292-5

[26] Davis, A, Coates, W, Stevens, K, The Post Office Experience: Designing a Large Asynchronous Chip. In 26th Hawaii International Conference on System Science, pp 409-416 (January 1993).

[27] Dobberpuhl, DW, et al, A 200 MHz 64-bit dual-issue CMOS microprocessor. Digital Technical Journal 4(4), pp 33-50 (1993).

[28] Endecott, PE, Furber, SB, Behavioural Modelling of Asynchronous Systems for Power and Performance Analysis. PATMOS'98 (1998).

[29] Endecott, PE, LARD Documentation. URL: http:/www.cs.man.ac.uk/amulet/ projects/lard

[30] EPIC TimeMill.

[31] Fant, KM, Brandt, SA, Null Convention Logic TM. Theseus Logic Inc. (1997). URL: http://www.theseus.com/Downloads/NCLPaper.pdf

[32] Ferranti Sales Literature. Universal High-Speed Digital Computers: A Small Scale Experimental Machine (August 1952). URL http://www.computer50.org/kgill/ mark1/sale/html.

[33] Furber, SB, Garside, JD, Riocreux, PA, Temple, S, Day, P, Liu, J, Paver, NC, AMULET2e: An Asynchronous Embedded Controller. Proceedings of the IEEE 87(2), pp 243-256 (February 1999).

[34] FUTUREBUS : Specifications for Advanced Microcomputer Backplane Buses, IEEE Computer Society Press (November 1983).

[35] Garside, JD, Bainbridge, WJ, Bardsley, A, Clark, DM, Edwards, DA, Furber, SB, Lui, J, Lloyd, DW, Mohammadi, S, Pepper, JS, Petlin, O, Temple, S, Woods, JV, AMULET3i - An Asynchronous System-on-Chip In Sixth International Symposium on Asynchronous Circuits and Systems, ASYNC2000, IEEE Computer Society Press (April 2000).

[36] Gilbert, DA, Dependency and Exception Handling in an Asynchronous Microprocessor. Ph.D. thesis, Department of Computer Science, The University of Manchester, UK (December 1997).

[37] Gilbert, DA, Garside, JD, A Result Forwarding Mechanism for Asynchronous Pipelined Systems. In Third International Symposium on Advanced Research in Asynchronous Circuits and Systems, ASYNC'97. Department of Computer Science, The University of Manchester (April 1997).

[38] Gwennap, L, ARM10 Points to Set-Tops, Handhelds, Microprocessor Report pp14-16, 12(15), (November 1998).

[39] Halfhill, TR, PowerPC 405GP has CoreConnect Bus, Microprocessor Report pp 8-9, 13(9), (July 1999).

[40] Hoare, CAR, Communicating Sequential Processes. Communications of the ACM 21(8), pp 666-677 (August 1978).

[41] Horstmann, JU, Eichel, W and Coates, RL, Metastability behavior of CMOS ASIC flip-flops in theory and test. IEEE Journal of Solid State Circuits 24(1), pp 146-157 (February 1989).

[42] HSPICE version 96.3, Meta-Software Inc, available from Avant!.

[43] Huffman, D.A., The synthesis of sequential switching circuits, J.Franklin Institute, (March/April 1954)

[44] IBM Corporation, PLB and OPB Macro Library, First Edition, (February 1999).

[45] IBM Corporation, CoreConnect Bus Architecture, product brief. URL: http://www.chips.ibm.com/news/1999/990923/pdf/corecon128_pb.pdf

[46] Intel Corporation. Intel Architecture Software Developers Manual. Intel Corporation, (1996).

[47] Intel Corporation, Pentium III Xeon Processor at 500 and 550 MHz Datasheet (February 1999).

[48] Jaggar, D, Advanced RISC Machines Architectural Reference Manual, Prentice Hall (1996).

[49] Josephs, MB, Yantchev, JT, CMOS Design of the Tree Arbiter Element. IEEE Transactions on Very Large Scale Integration (VLSI) Systems 4(4), pp 472-476 (December 1996).

[50] Kinniment, DJ, and Woods, JV, Synchronisation and arbitration circuits in digital systems. Proceedings of the IEE 123(10), pp 961-966 (October 1976).

[51] Kinniment, DJ, Yakovlev, AV, and Gao, B, Metastable Behaviour and System Performance. In Second UK Asynchronous Forum. S.B. Furber and A.V.Yakovlev, Editors. (July 1997).

[52] Lewis, M, Garside, JD, Brackenbury, L, Reconfigurable Latch Controllers for Low Power Asynchronous Circuits. In Fifth International Symposium on Advanced Research in Asynchronous Circuits and Systems, ASYNC'99. Department of Computer Science, The University of Manchester (April 1999).

[53] Liu, J, Arithmetic and Control Components for an Asynchronous System. Ph.D. thesis, Department of Computer Science, University of Manchester, UK (1997).

[54] Martin, AJ, Synthesis of Asynchronous VLSI Circuits, Department of Computer Science, California Institute of Technology (August 1991).

[55] Martin, AJ, The Design of a Self-timed Circuit for Distributed Mutual Exclusion. In 1985 Chapel Hill Conference on VLSI, pp 245-260 (1985).

[56] Martin, AJ, The limitations to delay-insensitivity in asynchronous circuits. In WJ. Dally, editor, Sixth MIT Conference on Advanced Research in VLSI, MIT Press, pp263-278 (1990).

[57] Matick, RE, Transmission Lines for Digital and Communication Networks, MGraw-Hill (1969).

[58] Molina, PA, The Design of a Delay-Insensitive Bus Architecture using Handshake Circuits. Ph.D. thesis, Imperial College of Science, Technology and Medicine, University of London, UK (1997).

[59] Molina, PA, Cheung, PYK, A Quasi Delay-Insensitive Bus Proposal for Asynchronous Systems. In Third International Symposium on Advanced Research in Asynchronous Circuits and Systems, ASYNC'97 (March 1997).

[60] Nanya, T, Takamura, A, Kuwakao, M, Imai, M, Fujii, T, Ozawa, M, Fukasaku, I, Ueno, Y, Okamoto, F, Fujimoto, H, Yamashina, M, Fukuma, M, TITAC-2: A 32-bit Scalable-Delay-Insensitive Microprocessor, HOT Chips IX, Stanford, pp 19-32 (August 1997).

[61] Nordholtz, P, Treytnar, D, Otterstedt, J, Grabinski, H, Niggemeyer, D, Williams, T, Signal Integrity Problems in Deep Submicron Arising from Interconnects between Cores., VLSI Test Symposium, Monterey (April 1998).

[62] Nowick, SM, Automatic Synthesis of Burst-Mode Asynchronous Controllers, Ph.D. thesis, Stanford University (1993).

[63] Nowick, SM, et al MIMIALIST: An Environment for the Synthesis, Verification and Testability of Burst-Mode Asynchronous Machines. Tecc. Report #CUCS-020-99, Columbia University Computer Science Dept. (July 1999)

[64] PI-Bus. Draft Standard, OMI324:PI-Bus Rev 0.3d, Open Microprocessor Systems Initiative (OMI), Siemens AG (1994).

[65] Paver, NC, The Design and Implementation of an Asynchronous Microprocessor. Ph.D. thesis, Department of Computer Science, University of Manchester, UK (1994).

[66] Peeters, AMG, Single-Rail Handshake Circuits. Ph.D. thesis, Technische Universiteit Eindhoven, Eindhoven, The Netherlands (1996).

[67] Peterson, JL, Petri-Net Theory and the Modelling of Systems, Prentice Hall Inc. N.J. (1981).

[68] Petlin, OA, Design for Testability of Asynchronous VLSI Circuits. Ph.D. thesis, Department of Computer Science, The University of Manchester, UK (1996).

[69] Renaudin, M, Vivet, P, Robin, F, ASPRO-216: a Standard-Cell Q.D.I. 16-Bit RISC Asynchronous Microprocessor. In Fourth International Symposium on Advanced Research in Asynchronous Circuits and Systems, ASYNC'98. ENST de Bretagne, France (March 1998).

[70] EIA-RS232 Interface between data terminal equipment and data circuit terminating equipment employing serial binary data interchange, Electronic Industries Association (revision F, 1997).

[71] Seitz, C, "System Timing", Chapter 7 of Introduction to VLSI Systems by Mead, C, Conway, L, Addison Wesley. Second Edition (1980).

[72] Shanley, T, Anderson, D, PCI System Architecture. ISBN 0-201-40993-3. Addison-Wesley. Third Edition (1995).

[73] Sobelman, GE, Fant, K, CMOS Circuit Design of Threshold Gates with Hysteresis, Theseus Logic Inc.,
URL: http://www.theseus.com/Downloads/ISCAS98P.pdf

[74] Stevens, KS, Practical Verification and Synthesis of Low Latency Asynchronous Systems. Ph.D. thesis, University of Calgary, Alberta, Canada (September 1994).

[75] Sutherland, IE, "Micropipelines", Communications of the ACM, **32** (6), pp 720-738 (June 1989)

[76] Sutherland, IE, Molnar, CE, Sproull, RF, Mudge, JC, The Trimosbus. In Proceedings of the First Caltech Conference on Very Large Scale Integration, CL Seitz, Editor (1979).

[77] Small Computer System Interface (SCSI), American National Standards Institution, (1986).

[78] Takamura, A, Kuwakao, M, Imai, M, Fujii, T, Ozawa, M, Fukasaku, I, Ueno, Y, Nanya, T, TITAC-2: A 32-bit Scalable-Delay-Insensitive Microprocessor, Proceedings of ICCD'97, pp 288-294 (October 1997).

[79] The National Technology Roadmap for Semconductors Technology, Semiconductor Industry Association (SIA),(1997).

[80] van Berkel, K, Beware the Isochronic Fork, Integration 13(2), pp103-128 (June 1992).

[81] van Berkel, K, Handshake Circuits, An asynchronous architecture for VLSI programming. Cambridge International Series on Parallel Computation. (1993).

[82] van Berkel, K, Kessels, J, Roncken, M, Saeijs, R, Frits, S, The VLSI-Programming Language Tangram and its translation into Handshake Circuits. In European Conference on Design Automation, pp 384-389 (1991).

[83] van Gageldonk, H, An Asynchronous Low-Power 80C51 Microcontroller. Ph.D. thesis, Eindhoven University of Technology, The Netherlands (1998).

[84] Verhoeff, T, Encyclopedia of Delay-Insensitive Systems, Eindhoven University of Technology, The Netherlands (1995-1998)
URL: http://edis.win.tue.nl/edis.html

[85] Verhoeff, T, Peeters, AMG, The Asynchronous Bibliography, Eindhoven University of Technology, The Netherlands (1991-1999).
URL http://www.win.tue.nl/cs/pa/wsinap/async.html

[86] VLSI Technology Inc. 0.35 micron, 3-layer metal CMOS process (VCMN4A3)

[87] VMEbus Specification Manual, VMEbus Manufacturers Group (1982).

[88] Weste, NHE, Estraghan, K, Principles of CMOS VLSI Design, A Systems Perspective. Addison Wesley. Second Edition (1993).

[89] Williams, FC, Kilburn, T, Electronic Digital Computers. Nature 162, pp 487 (September 1948). URL http://www.computer50.org

[90] Williams, FC, Kilburn, T, Toothill, GC, Universal High-Speed Digital Computers: A Small Scale Experimental Machine. In Proceedings of the IEE, pp 487 (February 1951). URL http://www.computer50.org

Index